To: JOHN DORSEY

MY OLD FRIEND
& THE BEST DAMN SAILOR
I KNOW!!

Jim Thompson
16 OCTOBER, 2000

Decorations, Medals, Ribbons, Badges and Insignia of the United States Navy

World War II to Present

By
James G. Thompson

1st Edition

This book is respectfully dedicated to:

RMT 1C Thomas Willfred Warren
United States Naval Reserve

and the millions of Navy Veterans, who have served our Country

"Fair Winds and Following Seas!"

Library of Congress Catalog Card Number - 99-074241
Hardcover Edition ISBN - 1-884452-50-7
Softcover Edition ISBN - 1-884452-51-5

Published by:

MOA Press (Medals of America Press)
114 South Chase Blvd.
Fountain Inn, SC 29644
Telephone: (864) 862-6051
www.moapress.com
www.usmedals.com

Printed in the United States of America

About the Author

James G. "Jim" Thompson

James G. "Jim" Thompson, attended the University of Wisconsin-Madison on a Naval ROTC scholarship, earning Bachelor's Degrees in both Economics and Naval Science. Following graduation, he served as a regular officer in the United States Marine Corps from 1960 to 1964. His service included time as a Platoon Commander and Assistant S-4 in the 2nd Battalion, 2nd Marine Regiment, as well as Assistant S-4 for Embarkation on the staff of the 2nd Marines. He served with the Navy as the Executive Officer of the Marine Detachment aboard the USS Randolph (CVS-15). Mr. Thompson also sailed on the USS Wisconsin (BB-64) as a Midshipman and the USS Monrovia (APA-31) and the USS Okinawa (LPH-3) as an officer in the Fleet Marine Force. He is a retired Sales Manager formerly with Procter & Gamble and resides with his wife Hannelore in Dunwoody, Georgia.

This is Mr. Thompson's second book; his first book, ***"THE DECORATIONS, MEDALS, RIBBONS, BADGES AND INSIGNIA OF THE UNITED STATES MARINE CORPS"*** is also available from Medals of America Press.

Grateful Acknowledgments

Secretary of the Navy
Commander J. Mahar, USN, *Secretary, Navy Department Board of Decorations and Medals*

Chief of Naval Operations - Navy Uniform Matters
Master Chief Petty Officer M. C. Cruse, USN,
Ms. Trudy Allen,

United States Marine Corps - Awards Branch
Colonel Fred Anthony, USMC (Ret.), *Former Director*
Mrs. Charlene Rose, *Former Assistant Director*

Institute of Heraldry, United States Army
Colonel Thomas B. Profitt, USA(Ret.), *Former Director*

Medals of America, Inc.
Mrs. Linda Foster, *President*
Colonel Frank Foster, USA (Ret.)
Mrs. Bonnie Crocker
Master Chief Petty Officer T.G. Dantzler, USN (Ret.)

Vanguard
Mr. Michael Harrison, *Vice-President*
Mr. Gary Duncan, *General Manager*

Other
Mr. Lawrence H. Borts, *Author and Consultant*
Mrs. Hannelore Thompson,
Supportive wife and diligent proof reader

A very special thank you to Mr. John A. Stacey, former Marine, author and historian, for the use of his research on Navy Specialty and Distinguishing Marks.

Mrs. Shelby Jean Kirk
Branch Head (1978-1999)
Office of Naval Operations
Awards & Special Projects Branch

"Over the last 31 years I have seen awards created and decimated, rules and regulations expanded and contracted, procedures and policies systemized and simplified. I have received messages; phone calls, facsimiles, and letters from every ocean the Navy sails, every rank of Sailor, and every level of command. I have worked personally on many occasions for the last nine CNOs and have seen many changes in the philosophy of awards by the leadership and, in turn, the regard for the awards by the fleet.

Three truths have been evident. First, the system is built on trust; a trust that the medal means the same as it did thirty years ago, the same from coast to coast. Second, the awarding authority and awards board cannot allow popular causes and senior influence to interfere or re-interpret the definition of the award. And finally, the scrutiny that often times delays the award is necessary to ensure that the award will stand up over the course of time and the criticism of peers. I love the Navy's Awards Program and have labored many a day with it; I have encouraged Jim Thompson in the writing of this book to ensure that it was respectful and accurate. I believe it is both and commend Jim on a job well done."

— Mrs. Shelby Jean Kirk ⚓

Table of Contents

Introduction 5
Background and History 6
Navy Insignia 11
Ranks, Rates and Ratings 13
Officer Rank Insignia 14
Commissioned Officer Sleeve Devices 17
Warrant Officers Sleeve Devices 18
Enlisted Rank/Rate Insignia 20
Service Stripes 22
Specialty Marks 23
Distinguishing Marks 32
Breast Insignia 35
Identification Badges 40
Aiguillettes, Brassards and Buttons 44
Awards and Decorations 46
Types of Medals and Ribbons 47
Attachments and Devices 48
Placement of Devices 50
Claiming Medals 53
Wearing Ribbons and Medals 49, 54
Wearing Medals, Insignia and the Uniform by Veterans and Retirees 55
Color Plates 56
Decorations and Service Medals 74
Foreign Decorations 105
Marksmanship Badges 115
Commemorative Medals 116
Displaying Awards 118
Bibliography 120
Index 121

List of Illustrations

Placement of Devices on Ribbons and Medals 50
Award Certificates 52
Uniform Insignia 57
Breast Insignia 58
Identification Badges 61
Rank and Rate Insignia 62
Line and Staff Corps Devices 63
Shoulder Boards, Rating Badges and Service Stripes 64
Medals of Honor 65
U.S. Personal Decorations 66
U.S. Personal Decorations and Service Medals 67
U.S. Service Medals 68
U.S. Service Medals and Foreign Decorations 69
Foreign Decorations, Non-U.S. Service Awards and Marksmanship Awards 70
Examples of Award Displays 71
Ribbon Precedence Chart 72
Medal and Ribbon Devices 73

Several years ago, while contemplating retirement from a career of 32 years with Procter & Gamble, I was approached by an old friend who is in the publishing business, with the suggestion that I write reference books on the badges and decorations of the Navy and the Marine Corps. At first I was puzzled as to why he had asked me, even though he knew that I had been commissioned through the NROTC (Regular) Program and served for four years as a regular officer in the Marine Corps. His intent was to convince me of the fact that there was a need for such a book should be sufficient reason to take on the job. Still doubtful, I argued that there must be many books available on this subject, so why should I do a "me too" book. He asked me to prove the validity of my objection to myself by researching ***"Bowker's Books in Print"*** and ***"Baker and Taylor"*** to find out what books were currently available on the subject. Following his suggestion, I learned, to my surprise, that although there were books about military insignia and books on U.S. decorations, there was little or nothing in print on these subjects devoted specifically to the Navy and the Marine Corps. With this realization and a little more coaxing I decided to write this book and one on the Marine Corps.

This book is designed to be a definitive reference covering Navy decorations, medals, ribbons, badges and insignia since World War II. The book is written for the Navy: Naval personnel on active duty, veterans, and those who are interested in the Navy. The book is intended to be a single source on these subjects and detailed enough to meet the needs of historians and collectors. Material for this book was derived from many authoritative sources, but the most valuable information came from the Navy itself: Naval personnel, The Navy Awards and Special Projects Branch and The Navy Uniform Matters Office. Portions of this book were taken directly from the current ***"Navy Uniform Regulations"*** and ***"All Hands"*** to insure accuracy.

This book, like all reference books, is out of date almost as soon as it is published. In the area of decorations, changes are being made constantly. For example, there have been eighty-four new U.S. awards created since 1947, or an average of over $1^1/_2$ new awards a year. Although the Naval Service has had fewer changes than other services since World War II, it too is difficult to stay current.

I realize that in writing this book, I open myself to the close scrutiny and criticism of those on active duty, reservists, veterans, historians and collectors. As with any book, there are bound to be mistakes. If errors are detected, please accept my apologies and the assurance that they will be rectified in future editions. Factual errors may also be found in this book and I encourage those of you seeking the ultimate in accuracy to communicate corrections to me through the publisher, or directly.

The history and tradition of the United States Navy, its decorations, medals, ribbons, badges and insignia are a part of what makes the Navy what it is. The United States Navy has a rich tradition of symbols; these bits of metal, cloth and ribbon add special character, and this book, in a small way, is intended to provide a good reference to this legacy.

Jim Thompson
Dunwoody, Georgia

Specific questions on information provided in this book should be addressed to:

Uniforms
Navy Uniform Matters
Navy Annex Room 1055
Washington, DC 20370

CNO Awards
CNO N09B33
2000 Navy Pentagon
Washington, DC 20350-2000

SECNAV Awards
SECNAV NDBDM
1000 Navy Pentagon
Washington, DC 20350-1000

Background and History

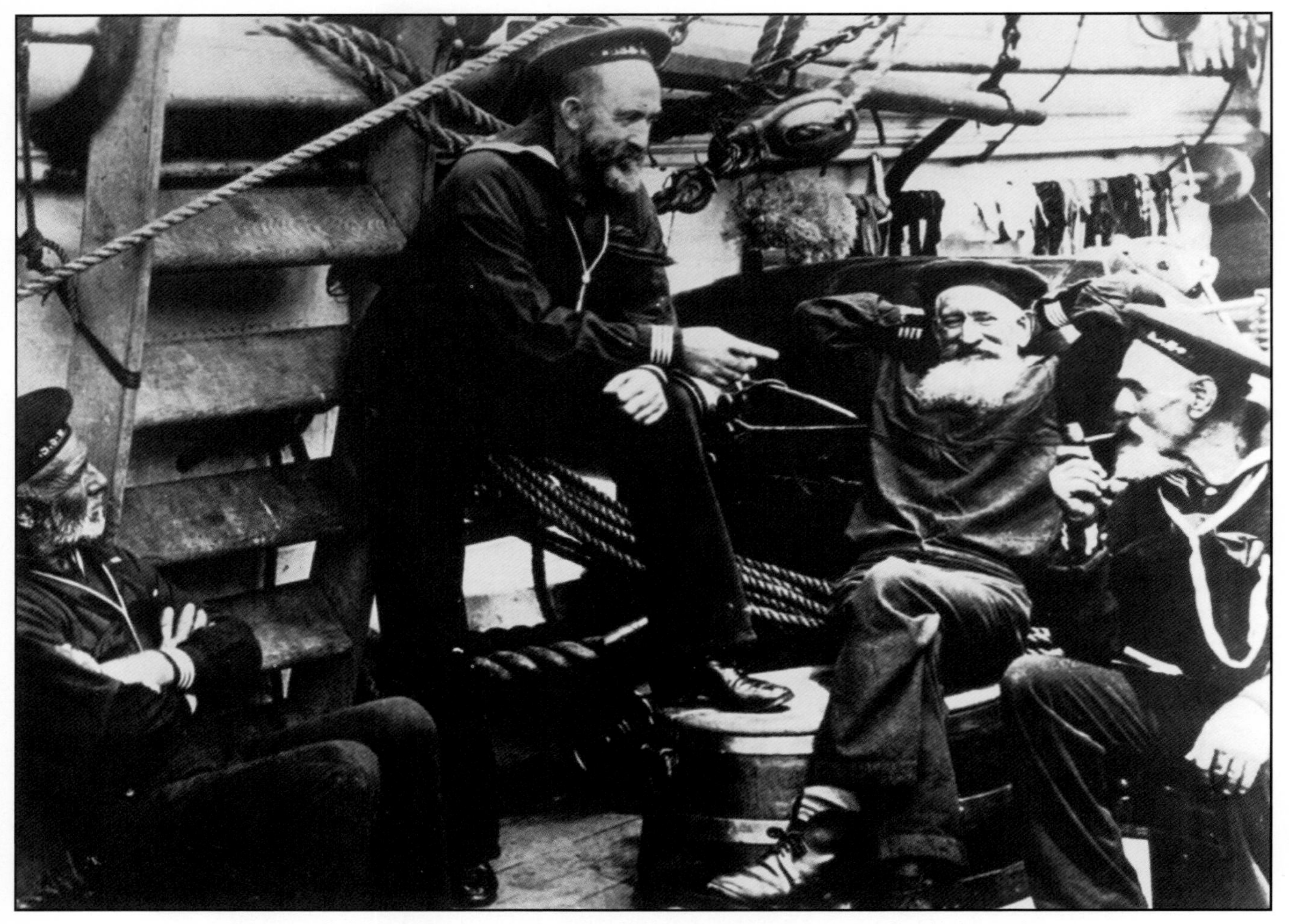

This post Civil War photograph shows Petty Officers with flat hats and cuffs with four stripes.

Uniforms - The first uniform instructions for the United States Navy were issued by the Secretary of War on 27 August 1791. This first set of instructions provided for a distinctive dress for officers who would command ships of the Federal Navy. The instructions didn't specify a uniform for enlisted personnel, but the usual dress for seamen was made up of a short jacket, shirt, vest, long trousers and a black crowned hat. The bell bottom trousers were introduced in 1817 and allowed men to roll them above the knee when washing down the decks. These first trousers had seven buttons until the early 1800's, when they had 15. (The 13 buttons common during World War II are said to be for the 13 original colonies, but there is no basis in fact for this.) The short jacket, or jumper, had a flap which was intended to protect it from the grease or powder normally used by seamen to hold their hair in place.

Neckerchiefs, or bandannas, were originally used as sweat bands and collar closures. The predominant color was black, which was less likely to show dirt.

The blue flat hat was first authorized in 1852 and vessel names were added in 1866 (a white sennit straw hat was authorized as an additional uniform item). The vessel names were removed in 1941 and replaced with U.S. NAVY for security reasons. Flat hats were finally eliminated in 1963. During the 1880's, the first white rolled brim hat appeared (which replaced the straw hat). This early canvas hat, with some modifications, became the white hat of today.

Flat Hat

White Hat

Stripes and stars were added to jumpers of enlisted personnel in 1866. This addition provided for three stripes for all grades on the collars and stripes denoting non-rated grade on the cuffs (one stripe for E-1, two for E-2, three for E-3 and above (petty officers were authorized four cuff stripes from 1869 to 1886).

Dungarees were introduced in 1901 and 1913 regulations originally permitted this uniform to be worn by officers and enlisted personnel with the cap of the day.

The first uniform for enlisted women was comprised of a single breasted coat (blue in winter and white in summer), long gull-bottomed skirts and a straight-brimmed hat (blue felt in the winter and white straw in the summer), black shoes and stockings. In the early days havelocks were also issued to female personnel, which was a protective cover worn over the combination cap to provide cold weather protection.

Khakis were introduced into the Navy in 1912 for naval aviators and submarine officers in 1931. During World War II, khakis were authorized for on-station wear by all officers and chiefs and finally for liberty as well. Gray working uniforms (the same style as khaki) were introduced in 1943 as a uniform for officers and chiefs. Prior to its discontinuation in 1949, the Navy "grays," were also authorized for cooks and stewards.

In 1917, aviation green uniforms were authorized for aviation officers as a winter working uniform. Chief petty officers were permitted to wear this uniform in 1941 when they were designated Naval Aviation Pilots. Women joined the aviation community in 1985 (but were not authorized the green uniform until later).

World War II saw the end of the cocked hat, which was also called the "fore and aft" hat. This formal hat was worn in the 1700's parallel to the shoulders and later (1800's to 1940) modified to be worn with the points to the front and back.

Badges and Insignia - The rich history of the United States Navy is expressed in many ways, but particularly in the badges and insignia worn over the years. As with the many of the traditions of our Navy, many of the badges and insignia stem from those seen in the British Navy. The foul anchor (customarily called "fouled" anchor), as an example, can be directly attributed to the influence of British Naval tradition.

Civil War Petty Officer's Badge

In 1841, insignia called "marks of distinction" were first prescribed as part of the official uniform. An eagle and anchor emblem, forerunner of the rating badge, is considered the first distinguishing mark (line petty officer's insignia had a five pointed star above the eagle). Petty officers wore this badge from 1841 until 1886, when rating badges began showing 15 specialty marks provided to cover the various ratings of that period.

The Civil War also saw the first flag rank, Commodore, established on 16 July 1862. The Commodore grade has remained until this day, but the title was changed to Rear

Commodore/Rear Admiral (Lower Half)

Admiral in 1900. The title Commodore was then re-instated for World War II and then again in 1980. Currently, this first flag rank title is Rear Admiral (Lower Half), which was established in 1985.

Officers stars were first approved on line officers' uniforms on 28 January 1864. The star points downward toward the gold stripe on the sleeve according to regulations since 1873.

Rating Badge - 1886 (Quartermaster)

Beginning in 1866, a series of marks designed to distinguish between petty officers by function (rating) were introduced to be worn in addition to the "petty officer's device." These marks continued in use, with some additions and modifications, until the combination rating badge was introduced in 1886. The combination rating badge distinguished both job specialty and grade and included an eagle, a specialty mark and chevron(s).

The early eagle and specialty marks were blue on the white uniform and white on the blue uniform, with chevrons of scarlet for both uniforms (blue chevrons on whites came later). The early regulations also provided that petty officers holding three consecutive good conduct medals would have chevrons in gold lace instead of scarlet cloth. In the early years these rating badges were worn on the right sleeve by petty officers of the starboard watch and the left sleeve for those of the port watch. Non-rated men wore a 3/8 inch stripe around the sleeve seam of the jumper, which was referred to as a watch mark, and was worn until 1912, when they were renamed. The new marks, called branch marks, were worn by enlisted men until they earned a "rating." These marks were white on the blue uniform and blue on the white uniform for seamen and scarlet on the uniforms for those in the engineering ratings.

Seaman **Fireman**

Distinguishing Marks

The term "distinguishing mark" was introduced in Uniform Regulations of 1905. These marks were defined as sleeve markings for those who had met certain qualifications

in addition to those required for their rating, or who were members of a crew that had attained a special merit in certain required competitions. During World War II, distinguishing marks were worn mid-way between the wrist and the elbow on the right sleeve for men of the seaman branch, and the left sleeve for all others. Distinguishing marks, although now obsolete, are the forerunners of the breast insignia worn today.

In 1913, the wearing of rating badges no longer corresponded to the assigned watch. Instead, men of the seaman branch wore the badge on the right sleeve and all others on the left sleeve. In 1941, all rating badges were redesigned so that the eagle faced to its left for petty officers in the seaman branch and to the right for all others; this way the eagle would face to the wearer's front. It should also be mentioned that, during this time, petty officers assigned to Fleet Marine Force units began wearing rating badges on Marine uniforms. This same type of rating badge was also worn on green uniforms by members of Seabee units and chief petty officers with aviation units.

Rating Badges

The 24 February change to the 1947 Uniform Regulations eliminated "right arm ratings" for the seaman branch and all rating badges would be worn on the left sleeve with the eagle facing the wearer's front. With this change came the elimination of the branch marks on the shoulder seam and the use of cuff stripes (one, two and three stipes) for non-rated personnel. These white cuff stripes, three for all enlisted men below the rate of chief petty officer, were retained as decorative trim on blue dress jumpers. These changes ultimately became effective in April 1948.

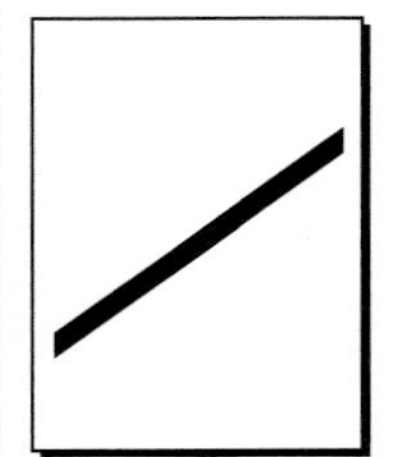

Group Rating Marks

During World War II WAVES wore the first group rating marks and in 1948 they were authorized for male personnel below the rate of petty officer. These marks were 3 inch angled stripes (1/4 inch wide and 1/4 inch apart). These marks, titled group rate marks, were blue on the white uniforms and white on the blue uniforms; for seaman (3 stripes), seaman apprentice (two stripes) and seaman recruit (one stripe); hospitalman and hospitalman apprentice (the latter two with a caduceus above the stripes); dentalman and dental apprentice (all with a caduceus with superimposed "D" above the stripes); and stewardsman and steward apprentice (these two had a steward's crescent mark above the stripes). The marks were scarlet on all uniforms for fireman and fireman apprentice. Emerald green was selected for all uniforms for airman and airman apprentice. Constructionman, construction apprentice marks were light blue. Personnel above recruit who qualified and were designated strikers, were permitted to wear their specialty mark above the stripes. Regulations state that these strikers marks are to be centered above the rectangular backing of the group rate.

Dungaree Rating Badge (Petty Officer 1st Class)

In 1951, Uniform Regulations introduced a dungaree rating badge in black printed on the a blue chambray material for wear with the blue work shirt. These badges were made to be ironed on the shirt and did not have specialty marks. Current petty officer rating badges are blue on the white uniforms and white on the blue uniforms, with blue chevrons on the white uniforms and scarlet chevrons on the blue uniforms.

The first qualification or skill badge was the naval aviator badge which has the American Shield in the center superimposed over a fouled anchor with a pair of wings bearing the shield and anchor. This gold insignia has been in effect since the start of naval aviation and was originally struck in 14-carat gold.

Naval Aviator Wings

The command at sea pin was introduced in 1960 to recognize the responsibilities placed on those officers of the Navy who are in command of, or have successfully commanded, ships and aircraft squadrons of the fleet. The badge is worn as a breast insignia on the right breast when in command and on the left breast after the command assignment is terminated.

Command at Sea Pin

1975 uniform regulations omitted the group rate mark for the recruit classification and stipulated that the E-1 pay grade is not required to wear the group rate marks.

Over the years there have been many changes to the badges and insignia and each one adds a colorful chapter to the history of the Navy.

John Paul Jones Medal

1782 Badge of Military Merit   1932 Purple Heart

Decorations, Medals and Ribbons - Napoleon wrote, "A soldier will fight long and hard for a bit of colored ribbon". Wellington, Napoleon's conqueror, introduced campaign medals to the British Army and the first went to troops who defeated Napoleon at Waterloo. Both Napoleon and Wellington realized that decorations and medals express national gratitude and stimulate esprit de corps.

The history of military decorations in the United States began early in the American Revolution when Congress voted to award gold medals to outstanding military leaders. The first medal was struck to honor George Washington for his service in driving the British from Boston in 1776. Similar medals were awarded to General Horatio Gates for his victory at the Battle of Saratoga and Captain John Paul Jones after his famous naval engagement with the Serapis in 1779. Unlike present practice, however, these were large presentation medals not designed to be worn on a uniform.

Andre Medal

In 1780, the Congress of the United States created the Andre Medals, which, for the first time, broke the custom of restricting the awarding of medals to senior officers. It is doubly unique in that it was designed to be worn with the uniform as a neck decoration. The medal was awarded to three enlisted men who captured Major John Andre who had the plans of the West Point fortifications in his boot.

In 1782, George Washington established the Badge of Military Merit, the first U.S. decoration that had general application to all enlisted men and was the forerunner of the Purple Heart. Washington hoped that this would inaugurate a permanent awards system. Although special and commemorative medals had been awarded previously, until this point no decoration had been established which honored the private soldier with an award of special merit. The object of the Badge of Military Merit was "to foster and encourage every species of military merit." The medal was a heart of purple cloth or silk, edged with narrow lace or binding. Unfortunately, the award fell into disuse after the Revolution and disappeared for 150 years.

On the 200th Anniversary of Washington's birth, 22 February 1932, the Badge of Military Merit was reborn as the Purple Heart. The Purple Heart took the heart shape of the earlier Badge of Military Merit with Washington's profile on a purple background. The words "For Military Merit" appear on the reverse in reference to its predecessor.

During the Civil War, the Medal of Honor was established and remained the only American military award until the Navy and Marine Corps authorized their Good Conduct Medals in the late 19th Century. It was not until the eve of the 20th Century that seven medals were authorized to commemorate the events surrounding the Spanish American War. One of these medals was to commemorate the victory of the naval forces under Commodore Dewey over the Spanish fleet at Manila Bay. This medal was awarded to all officers and enlisted personnel present during the expedition and became the country's first campaign medal.

1861 Navy Medal of Honor

When Theodore Roosevelt became President, he legislated the creation of medals to honor all those who had served in previous conflicts. By 1908, the United States had authorized campaign medals, some retroactive, for the Civil War, the Indian Wars, the Spanish American War, the Philippine Insurrection and the China Relief Expedition of 1900-01. The Services used the same ribbon, but different medals were struck for the Army and Navy. The custom of wearing the ribbons of the medals on a ribbon bar began during this period. The Army and the Navy used different precedence for wearing these ribbons, which established an independence in the creation and wearing of awards by each service that remains to this day.

At the time of the U.S. entry into World War I, the Medal of Honor, Certificate of Merit and the Navy/Marine Good Conduct Medals were the only personal decorations. In 1918, the Army's Distinguished Service Cross and Distinguished Service Medal were established. That same year, the law, which prevented individuals from accepting foreign decorations, was rescinded. In 1919, the Navy created the Navy Cross and the

Army Distinguished Service Cross, Navy Cross and Navy Distinguished Service Medal

Navy Distinguished Service Medal for Navy and Marine personnel. The issuance of the World War I Victory Medal established the practice of wearing clasps with the names of battles on the suspension ribbon, which was a practice in many countries. When the ribbon bar was worn, a bronze star represented each clasp.

American, Asiatic-Pacific and European-African-Middle Eastern Campaign Medals

On 8 September 1939, President Franklin Roosevelt proclaimed a National Emergency and the first peacetime service award, the American Defense Service Medal, was established. At the beginning of World War II, the United States increased the number of both personal decorations as well as campaign medals. Since U.S. forces were serving all over the world, a campaign medal was established for each major theater. The three medals were American Campaign, Asiatic-Pacific Campaign and European-African-Middle Eastern Campaign. The World War I practice of using clasps to denote campaigns on the suspension ribbon was discarded in favor of the three-sixteenth inch bronze stars.

UN Korean Service, U.S. Service Medals and National Defense Service Medals

Following World War II, the World War II Victory Medal and the Occupation Medals (for both Europe and Japan) were authorized. During the Korean Conflict the Korean Service Medal and the United Nations Service Medal were established along with the National Defense Service Medal. The National Defense Service Medal later became our country's most awarded medal when it was reinstated for the Vietnam and Gulf Wars.

U.S. Vietnam Service and RVN Campaign Medals

The first American advisors in the Republic of South Vietnam were awarded the new Armed Forces Expeditionary Medal, created in 1961, to cover campaigns for which no specific medal was instituted. As U.S. involvement in Southeast Asia grew, the Vietnam Service Medal was authorized. Uniquely, the previous recipients of the Expeditionary medal were given the opportunity to decide which award to accept. The Department of Defense also authorized the acceptance of the Republic of Vietnam Campaign Medal by all who served six months in-country or in the surrounding waters from July 1965 to March 1973.

During the Vietnam era and immediately following, the Department of Defense developed several new decorations including the Defense Distinguished Service Medal, the Defense Superior Service Medal and the Defense Meritorious Service Medal. Each of these awards was designed to recognize achievements of individuals assigned to the Office of the Secretary of Defense or other activities in the Department of Defense.

U.S. Southwest Asia Service Medal, Kuwait Liberation Medal (Saudi Arabia) and Kuwait Liberation Medal

The Gulf War saw the reinstatement of the National Defense Service Medal (this time it also included the Reserves) and the creation of the Southwest Asia Service Medal. The Department of Defense also approved the acceptance and wearing of the Kuwait Liberation Medals from Saudi Arabia and Kuwait.

During the 1990's many Naval personnel were awarded the NATO (North Atlantic Treaty Organization) Medal for service under NATO command or in direct support of NATO operations in the former Yugoslavia.

The most recent awards include the Navy Recruit Training Service Ribbon, established in 1998 and the Kosovo Campaign Medal established in 2000.

The insignia of the United States Navy is rich with history. The first "Navy" seal was adopted by the Continental Congress on 4 May 1780.

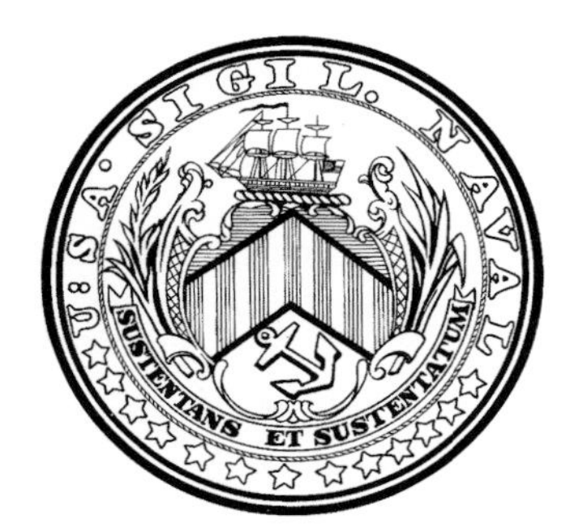

Board of Admiralty Seal

The Board of Admiralty Seal - is based on information found in Rough Journals of the Continental Congress. This representation was prepared at the request of the Secretary of the Navy for President John F. Kennedy. The seal is on a circular background, with a three masted square rigged ship underway, supported by a stylized sea scroll, over an inclined anchor. Below the anchor is a scroll with the Latin words SUSTENTANS ET SUSTENTATUM, which means "sustaining and having sustained," or "upholding and having upheld." The inscription around the edge is USA SIGIL. NAVAL at the top and thirteen stars around the bottom.

Seal of the Department of the Navy

The Seal of the Department of the Navy - is officially described as: "On a circular background of fair sky and moderate sea with land in sinister base, a three-masted square rigged ship under way before a fair breeze with after topsail furled, commission pennant atop the foremast, National Ensign atop the main, and the commodore's flag atop the mizzen. In front of the ship a Luce-type anchor inclined slightly bendwise with the crown resting on the land and, in front of the shank and in back of the dexter fluke, an American bald eagle rising to sinister regarding to dexter, one foot on the ground, the other resting on the anchor near the shank; all in proper colors. The whole within a blue annulet bearing the inscription DEPARTMENT OF THE NAVY at the top and UNITED STATES OF AMERICA at the bottom, separated on each side by a mullet and within a rim in the form of a rope, mullet and edges of annulet all gold."

The seal was made "official" on 23 October, 1957, but the central device is essentially the same used by the Department of the Navy when it was officially established on 30 April 1798.

Seal of the United States Navy

The Official Seal of the United States Navy - is officially described as: "Consists of the shield of the United States in front of an American bald eagle, wings spread, perched upon on a Luce-type anchor, which is displayed on a light background. The emblem is encircled with a navy blue band edged in a gold rope rim and inscribed UNITED STATES at the top and NAVY at the bottom, separated on each side by a mullet and within a rim in the form of a rope, mullet and edges of annulet all gold."

Officer and Chief Warrant Officer Cap Insignia

Officer and Chief Warrant Officer Cap Insignia - Two gold colored crossed fouled anchors with burnished silver colored shield surmounted by a burnished silver spread eagle facing to wearer's right (in 1941 the eagle was changed to face to the right; to the wearer's sword arm). It may be embroidered or made of metal (standard or high relief). This device is worn on the combination cap and a one-half size miniature version is worn on the garrison cap.

Warrant Officer, W-1 Cap Insignia (Obsolete)

Warrant Officer, W-1 Cap Insignia - Two crossed gold colored fouled anchors, of a size to be inscribed in a circle 2-1/4 inches in diameter. The unfouled arm of the stock points inward. It may be embroidered or made of metal. This device was worn on the combination cap and a one-half size miniature device was worn on the garrison cap.

Navy Nurse Corps Cap Insignia (Obsolete)

Navy Nurse Corps (obsolete) - An oak leaf with the initials NNC, superimposed on a gold colored fouled anchor (this insignia was worn during World War II). Today, Navy nurses wear the commissioned officer's cap insignia.

Chief Petty Officer Cap Insignia

Chief Petty Officer Cap Insignia - A gold colored fouled anchor with silver block letters USN superimposed on the anchor's shank. One, two, or three silver stars are attached above the anchor's stock indicating Senior Chief, Master Chief, and Master Chief Petty Officer of the Navy. This device is worn on the combination cap and a one-half size miniature device is worn on the garrison cap.

Enlisted Personnel Cap Insignia

Enlisted Personnel Cap Insignia - An oxidized, silver, spread eagle, with oxidized silver block letters USN placed horizontally between the wing tips and centered above the eagle's head. This insignia was worn by all enlisted personnel 1975 to 1985 on the combination cap worn with the obsolete CPO type uniform. This insignia is now worn only by female personnel on the women's combination cap and beret.

U.S.N.

Cook and Steward Cap Device (Obsolete)

Cook and Steward Cap Device (Obsolete) - Metal block letters, U.S.N. worn by cooks and stewards during World War II.

Midshipman Cap Device

Midshipman Cap Device - A gold colored fouled anchor (also worn by Officer Candidates and Naval Aviation Cadets).

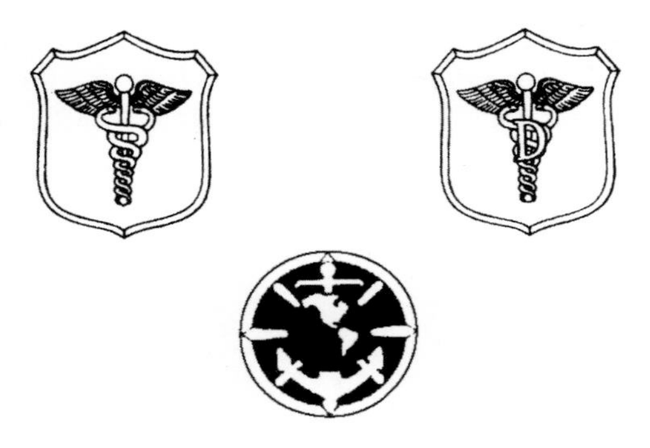

Hospital Corpsman, Dental Technician and Religious Program Insignia

Hospital Corpsman, Religious Program and Dental Technician Insignia - A non-glossy black insignia worn on the collars of utilities by Naval personnel in support of the Marine Corps.

Navy Retired Personnel Lapel Button

Navy Retired Personnel Lapel Button - A circular pin with a bald eagle, anchor and rays. Around the edge of the pin is a chain at the top and the raised inscription UNITED STATES NAVY at the bottom. The pin is silver for 20 years and gold for 30 years.

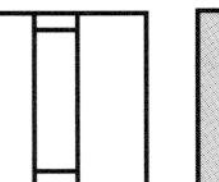
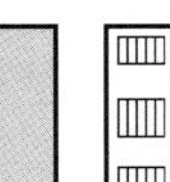

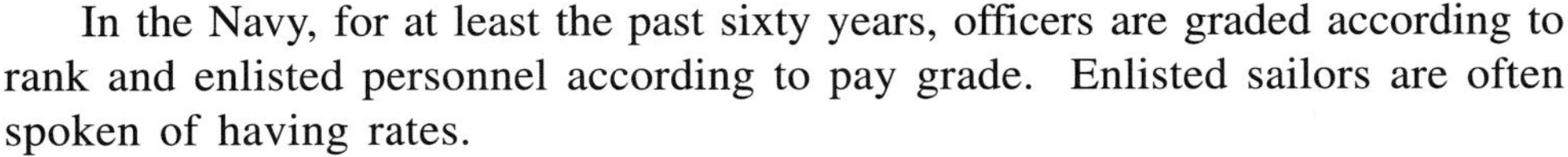
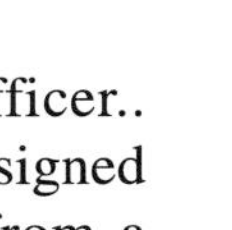

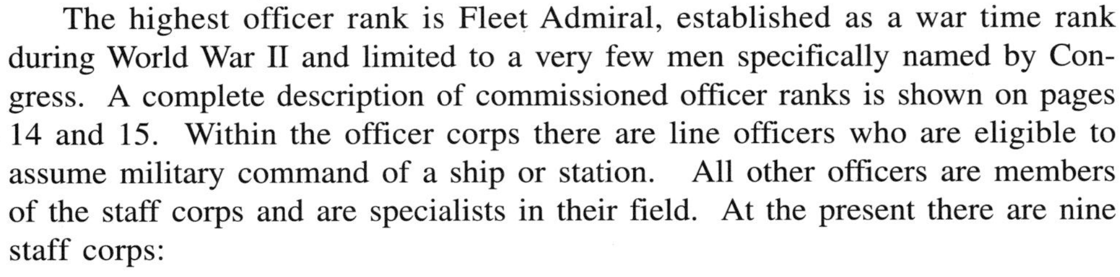

In the Navy, for at least the past sixty years, officers are graded according to rank and enlisted personnel according to pay grade. Enlisted sailors are often spoken of having rates.

A U.S. Naval officer may either be a commissioned or warrant officer.. Commissioned officers hold a commission granted by the President and signed by the Secretary of the Navy. Warrant Officers derive their authority from a warrant from the Secretary of the Navy.

The highest officer rank is Fleet Admiral, established as a war time rank during World War II and limited to a very few men specifically named by Congress. A complete description of commissioned officer ranks is shown on pages 14 and 15. Within the officer corps there are line officers who are eligible to assume military command of a ship or station. All other officers are members of the staff corps and are specialists in their field. At the present there are nine staff corps:

- Medical Corps
- Nurse Corps
- Supply Corps
- Chaplain Corps
- Civil Engineer Corps
- Dental Corps
- Judge Advocate General Corps
- Law Community
- Medical Service Corps

Line officers wear a star on their sleeves, or on the shoulder board above the stripes of rank. Staff officers wear their own unique insignia in lieu of the star and these insignia are shown on page 17.

Warrant officers earn special status by their ability and experience. Their uniform is similar to commissioned officer's with differences noted on page 15. Individually, chief warrant officers are referred to by the title of their specialty i.e. chief boatswain, chief machinist, etc.. Individuals with the rank of warrant officer (W-1), were referred to by the title boatswain, machinist, etc.*(the grade of warrant officer W-1 is currently not in use).* Warrant officer specialties are shown on pages 18 and 19.

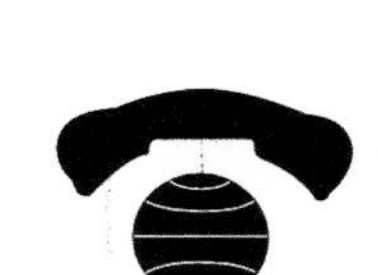

Midshipmen are generally classified as officers of the line in a qualified sense. Their uniforms are similar to commissioned officers', except only very thin gold striping is worn and gold anchor pin-on devices are worn on the collar. There are two types of midshipmen: Naval Academy midshipmen and midshipmen who are college students in the Naval Reserve Officer Training Corps (NROTC). Midshipmen rank below chief warrant officers.

Officers have rank, enlisted personnel have rates or ratings. (recruits are spoken of as non-rated men). A rating is a name given to an occupation in the Navy which requires certain aptitudes, training, experience, knowledge and skills.

Each rating has its own specialty mark which is worn on the left sleeve by all men properly qualified. Specialty marks are shown beginning on page 23. The rate is the pay grade (E-1 through E-9) which reflects a level of aptitudes, training, experience, knowledge, skill and responsibility. Pay grades are shown on page 20. The rating badge or insignia is a combination of the rate (pay grade) and rating (as indicated by the specialty mark above the chevrons or stripes).

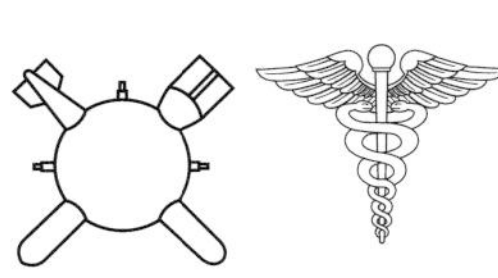

Officer Rank Insignia

Navy officers wear their rank devices in different places on their uniforms, depending upon the uniform. The three basic uniforms and the type of rank devices are: khakis (a working uniform) - pins on the collar; whites - stripes on shoulder boards; blues - stripes sewn on the lower sleeve. Shoulder boards are worn on summer service uniforms, white shirts, bridge coats and reefers. The collar devices are worn on the right side of the garrison cap (a miniature officer's insignia is worn on the left) and slightly larger devices are worn on the epaulets of the raincoat and working jacket.

Pay Grade	Rank	Abbreviation	Collar	Shoulder	Sleeve
O-11	Fleet Admiral	FADM			

FLEET ADMIRAL - Sleeve Insignia: One 2 inch stripe with four 1/2 inch stripes above it. Metal Grade Insignia: Five silver-colored, five-pointed stars. Shoulder stars are one inch in diameter and are either fastened together in the form of a circle. Collar stars are fastened together at the tips of adjacent rays in the form of a circle with each star pointing upward. *The rank of Fleet Admiral has been reserved for war time use only.*

Pay Grade	Rank	Abbreviation	Collar	Shoulder	Sleeve
O-10	Admiral	ADM			

ADMIRAL - Sleeve Insignia: One 2 inch stripe with three 1/2 inch stripes above it. Metal Grade Insignia: Four silver-colored, five-pointed stars. Shoulder stars are one inch in diameter and are either fastened together on a metal holding bar or placed individually with one point of each star in the same line; distance between the centers of adjacent stars is 3/4 inch. Collar stars are 9/16 inch in diameter and are fastened together on a metal holding bar in a straight line with one ray of each star pointing upward and at right angles to the holding bar.

Pay Grade	Rank	Abbreviation	Collar	Shoulder	Sleeve
O-9	Vice Admiral	VADM			

VICE ADMIRAL - Sleeve Insignia: One 2 inch stripe with two 1/2 inch stripes above it. Metal Grade Insignia: Three silver-colored stars, of the same type and arranged in the same manner as for a lieutenant general, except the distance between centers of adjacent shoulder stars is one inch.

Pay Grade	Rank	Abbreviation	Collar	Shoulder	Sleeve
O-8	Rear Admiral (upper half)	RADM			

REAR ADMIRAL - Sleeve Insignia: One 2 inch stripe with one 1/2 inch stripe above it. Metal Grade Insignia: Two silver-colored stars of the same type and arranged in the same manner as for a major general.

Pay Grade	Rank	Abbreviation	Collar	Shoulder	Sleeve
O-7	Rear Admiral (lower half)	RADM (LH)			

REAR ADMIRAL (Lower Half)/COMMODORE - Sleeve Insignia: One 2 inch stripe. Metal Grade Insignia: One silver-colored star (same type as above). *The title of commodor is not currently in use.*

Pay Grade	Rank	Abbreviation	Collar	Shoulder	Sleeve
O-6	Captain	CAPT			

CAPTAIN - Sleeve Insignia: Four 1/2 inch stripes. Metal Grade Insignia: A silver-colored spread eagle, worn in pairs, right and left talons of one foot grasping an olive branch, the other, a bundle of arrows. Shoulder insignia; slightly curved, with 1-1/2-inch wing span. Collar insignia: flat, with 31/32-inch wing span

Pay Grade	Rank	Abbreviation	Collar	Shoulder	Sleeve
O-5	Commander	CDR			

COMMANDER - Sleeve Insignia: Three 1/2 inch stripes. Metal Grade Insignia: A seven-pointed, silver-colored oak leaf, raised and veined. Shoulder insignia; slightly curved, one inch from stem tip to center leaf tip. Collar insignia: flat, 2-23/32 inch from stem tip to center leaf tip.

Pay Grade	Rank	Abbreviation	Collar	Shoulder	Sleeve
O-4	Lieutenant Commander	LCDR			

LIEUTENANT COMMANDER - Sleeve Insignia: Two 1/2 inch stripes with one 1/4 inch stripe in between. Metal Grade Insignia: A seven-pointed, silver-colored oak leaf, raised and veined. Shoulder insignia: slightly curved, one inch from stem tip to center leaf tip. Collar insignia: flat, 23/32 inch from stem tip to center leaf tip.

Pay Grade	Rank	Abbreviation	Collar	Shoulder	Sleeve
O-3	Lieutenant	LT	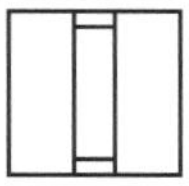		

LIEUTENANT - Sleeve Insignia: Two 1/2 inch stripes. Metal Grade Insignia: Two smooth silver-colored bars, without bevel, attached at each end by a holding bar. Shoulder insignia; each bar slightly curved, 1-1/8 inches long by 3/8 inch wide, and 3/8 inch apart. Collar insignia: flat, each bar 3/4 inch long by 1/4 inch wide and 1/4 inch apart.

Pay Grade	Rank	Abbreviation	Collar	Shoulder	Sleeve
O-2	Lieutenant Junior Grade	LTJG			

LIEUTENANT (JUNIOR GRADE) - Sleeve Grade Insignia: One 1/2 inch stripe with one 1/4 inch stripe above it. Metal Grade Insignia: One silver-colored bar of the same type as for a Lieutenant.

Pay Grade	Rank	Abbreviation	Collar	Shoulder	Sleeve
O-1	Ensign	ENS			

ENSIGN - Sleeve Insignia: One 1/2 inch stripe. Metal Grade Insignia: One gold-colored bar of the same type as for a 2nd lieutenant.

Pay Grade	Rank	Abbreviation	Collar	Shoulder	Sleeve
W-4	Chief Warrant Officer	CWO4			

CHIEF WARRANT OFFICER, W-4 - Sleeve Insignia: One 1/2 inch stripe with one break, centered on the outer face of the sleeve. Metal Grade Insignia: One silver-colored bar of the same type as for a Lieutenant (Junior grade), with three blue enamel blocks superimposed. Shoulder insignia; center enamel block is 1/4 inch wide, with 1/8 inch wide outer blocks, 1/4 inch from the edges of the center block. Collar insignia; center enamel block is 5/32 inch wide, with 3/32 inch wide outer blocks, 5/32 inch from the edges of the center block.

Pay Grade	Rank	Abbreviation	Collar	Shoulder	Sleeve
W-3	Chief Warrant Officer	CWO3			

CHIEF WARRANT OFFICER, W-3 - Sleeve Insignia: One 1/2 inch stripe with breaks 2 inches apart, two breaks centered symmetrically on outer face of sleeve. Metal Grade Insignia: One silver-colored bar of the same type as for a CWO-4, with two blue enamel blocks superimposed. Shoulder insignia blocks are 3/8 inch wide and 1/4 inch apart. Collar insignia; blocks are 1/4 inch wide and 5/32 inch apart.

Pay Grade	Rank	Abbreviation	Collar	Shoulder	Sleeve
W-2	Chief Warrant Officer	CWO2			

CHIEF WARRANT OFFICER, W-2 - Sleeve Insignia: One 1/2 inch stripe with breaks 2 inches apart, three breaks centered symmetrically on outer face of sleeve. Metal Grade Insignia: One gold-colored bar of the same type as for an Ensign, with three blue enamel blocks arranged in the same manner as for a CWO-4.

Pay Grade	Rank	Abbreviation	Collar	Shoulder	Sleeve
W-1	Warrant Officer	WO1			

WARRANT OFFICER, W-1 - Sleeve Insignia: One 1/4 inch stripe with breaks 2 inches apart, three breaks centered symmetrically on outer face of sleeve. Metal Grade Insignia: One gold-colored bar of the same type as for a CWO-2, with two blue enamel blocks arranged in the same manner as for a CWO-3. *The grade of WO1 is no longer in use.*

Hard Shoulder Boards

Combination insignia indicating wearer's grade and line or staff corps are curved to fit the shoulder and secured at their inner ends by a gilt button. Officers wear hard shoulder boards on reefers, overcoats, and designated uniforms. Women wear men's shoulder boards when wearing the men's reefer and women's shoulder boards when wearing the women's reefer.

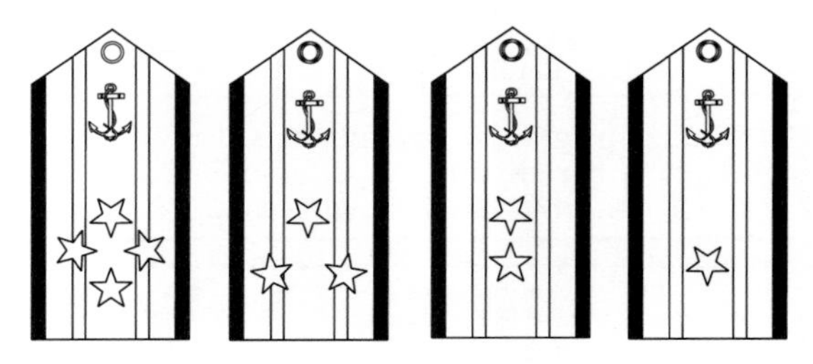

Flag Officers (Line) - The surface is covered with gold lace showing a 1/8 inch blue cloth margin on each of the long sides. A silver embroidered fouled anchor is placed with its center line along the shoulder board's longer dimension and the crown pointing toward the squared end of the board. The unfouled arm of the stock points to the front of the wearer (right and left). Designation of grade consists of silver embroidered five-pointed star(s), placed between the crown of the anchor and the squared end of the shoulder board.

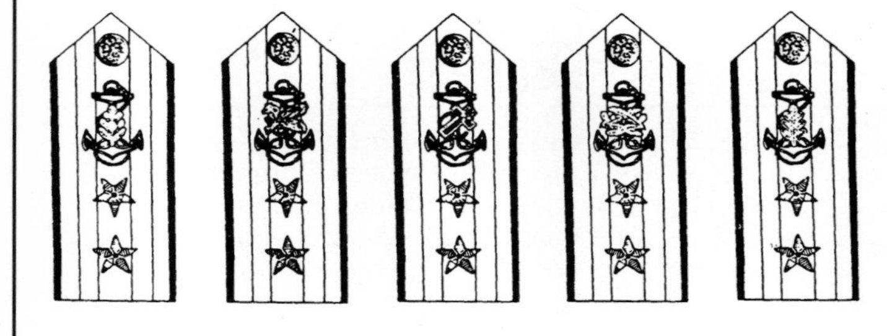

Flag Officer (Staff Corps) - The shoulder marks are the same as those described above with an appropriate corps device, of the same size as that prescribed for wear on the shirt collar, super-imposed on the anchor's shank.

Officers below Flag Grade - The surface is black cloth. Gold lace stripes, the same width, number, and spacing, specified for stripes on sleeves of the blue coat, designate rank. The first stripe starts 1/4 inch (1/2 inch for Ensigns) from the widest end. Line and staff corps devices replicate sleeve insignia and are placed as far from the stripes as specified for devices on sleeves of blue coats.

Warrant Officers - The surface is black cloth. Stripe widths are as specified for sleeves of blue coats but the blue break(s) are 1/2 inch rather than 2 inches apart. The stripe starts 1/2 inch from the widest end. Line or staff corps devices are sized as specified for devices on sleeves of blue coats.

Soft Shoulder Boards

Combination insignia indicating the wearer's grade and corps are 3/4 the size of the men's hard shoulder boards. Soft shoulder boards are worn on white epauletted shirts when worn with Service Dress Blue, and on Navy blue/black pullover sweater (wooly-pulley).

Metal Grade Insignia

Regular size metal grade insignia is centered on shoulder straps of blue all-weather coats/raincoats, khaki jackets, blue jackets and black jackets.

Collar grade insignia is worn on collar points. Line officers wear the grade insignia on both collar points. Staff Corps officers and chief warrant officers wear grade insignia on the right collar point and staff insignia on the left collar point. On the long sleeve khaki and blue shirt, the insignia is centered one inch from the front and upper edges of the collar (except for flag officers). Admirals, vice admirals, and rear admirals center the first star one inch from the front and upper edge of the collar, and position the vertical axis of the insignia at right angles (the horizontal axis parallel) to the upper edge of the collar. This procedure applies whether the collar is worn open or closed. On open collared short sleeve shirts the insignia is centered one inch from the front and lower edges of the collar and position the vertical axis of the insignia along an imaginary line bisecting the angle of the collar point.

Staff corps officers and warrant officers wear them on the left collar point of shirts in the manner described for collar grade insignia. The vertical axis of the insignia is aligned with the bisecting line of the collar point on open collar shirts, and the horizontal axis parallel to the upper edge of the collar on long sleeve khaki and blue shirts. They have the same design as those worn on the sleeves of blue coats but the dimensions conform to the specifications and standards sample. These devices are primarily in gold color metal with silver symbol as applicable. Leaves are veined, but otherwise smooth, and do not simulate embroidery.

Commissioned Officer Sleeve Devices

Commissioned officers are either line officers or staff corps officers. Those of the staff corps are specialists in career fields which are professions unto themselves, such as physicians, lawyers, civil engineers, etc. Staff corps officers wear their specialty insignia on the sleeve of the dress blue uniforms and on their shoulder boards in place of the star worn by line officers. On winter blue and khaki uniforms, the specialty insignia is a collar device worn on the left collar while the rank device is worn on the right.

Devices are embroidered, gold and silver as appropriate, or black on green uniforms and are worn as follows:

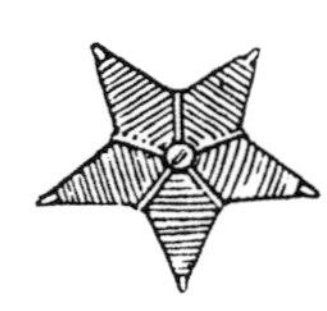

Line - Five-pointed gold star, with one ray pointing downward.

Medical Corps - Gold embroidered spread oak leaf, surcharged with a silver embroidered acorn, with stem down.

Nurse Corps - Gold embroidered spread oak leaf, sleeve, with stem down.

Supply Corps - Gold embroidered sprig of three oak leaves and three acorns, with the longer dimension parallel to the upper stripe, stem to the front (right and left). This insignia is also worn by Limited Duty Officers and Supply and Food Service Warrant Officers.

Chaplain Corps (Buddist) - Gold embroidered prayer wheel.

Chaplain Corps (Christian) - Gold embroidered, Latin cross, inclined to the rear; the longer arm makes an angle of 60 degrees with the upper stripe (right and left).

Chaplain Corps (Jewish) - Consists of the Star of David above and attached to the top center of the Tables of the Law, all in gold embroidery, with the shorter dimension parallel to the sleeve stripes.

Chaplain Corps (Muslim) - Gold embroidered crescent. Worn with moon opening to the front.

Civil Engineer Corps - Two overlapped gold embroidered sprigs of two live oak leaves, and a silver embroidered acorn in each sprig, with the longer dimension parallel to the sleeve stripes and the top pair opening to the front (right and left).

Dental Corps - Gold embroidered spread oak leaf, with a silver embroidered acorn on each side of the stem, with the longer dimension perpendicular to the sleeve stripes, stem down.

Judge Advocate General Corps - Two gold embroidered oak leaves, curving to form a semi-circle in the center of which is balanced a silver "mill rinde," with the longer dimension parallel to sleeve stripes, stems down.

Law Community - Silver, vertical mill rinde embroidered, superimposed on a silver quill pen and two gold oak leaves curve to form a semi-circle. Pen nib is down and left.

Medical Service Corps - Gold embroidered spread oak leaf, attached to a slanting twig, with the longer dimension perpendicular to the sleeve stripes, stem down, lower end of twig to the front (right and left).

Warrant Officer Sleeve Devices

Warrant officers have advanced through the enlisted ranks in various technical specialities and probabaly possess the most detailed and practical knowledge of the modern Navy. The insignia for the various warrant specialties are shown below:

Aerographer - Gold embroidered Aerographer - Gold embroidered device, consisting of a winged circle with a six feather arrow passing vertically through the circle, arrow pointing down and perpendicular to stripe, with the filled in half of the circle to the front (right and left).

Air Traffic Control Technician - Winged microphone.

Aviation Boatswain - Gold embroidered crossed winged anchors, with wings parallel to the stripe.

Aviation Electronics Technician - Winged, helium atom, embroidered in gold, with wings parallel to the stripe. The electron orbits diagonally with the bottom to the front (right and left).

Aviation Maintenance Technician - Winged, two-blade vertical propeller embroidered in gold, with wings parallel to the stripe.

Aviation Operations Technician - Two canted winged, crossed electron orbits, canted with a lightning bolt passing through toward wave lines, with wings parallel to the stripe (right and left).

Aviation Ordnance Technician - Gold embroidered, winged, flaming spherical shell, flame upward, parallel to the stripe.

Boatswain - Two gold embroidered crossed, fouled anchors, with crowns down and parallel to the stripe.

Communications Technician - Four gold embroidered lightning bolts, with the longer dimension parallel to the stripe, narrow end of device to the front (right and left).

Cryptologic Technician - A crossed quill nib to the front over a spark pointing down, with points down, parallel to the stripe (right and left).

Data Processing Technician - A gold embroidered quill, pen nib down to the front (right and left) superimposed diagonally on gear.

Diving Officer - Gold embroidered diver's helmet with face plate, square knot and breast plate bolts embroidered in silver.

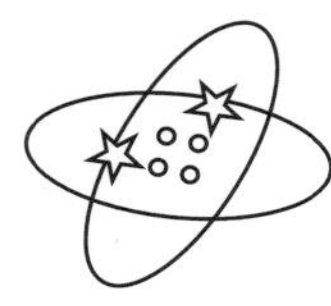

Electronics Technician - Gold embroidered helium atom, placed on the sleeve with the horizontal electron orbit parallel to the stripe. The electron orbits diagonally with the bottom to the front (right and left).

Engineering Technician - Gold embroidered, three-bladed propeller, with two blades down, and the lower edges parallel to the stripe.

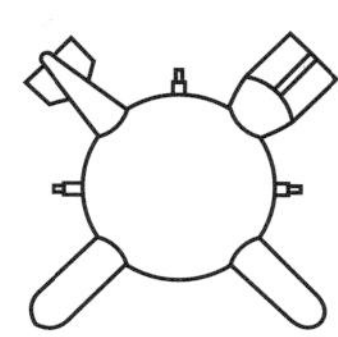

Explosive Ordnance Disposal Technician - A mine superimposed on a crossed torpedo and air craft bomb. The torpedo points down and aft; the bomb points down and forward (right and left).

Intelligence Technician - Gold embroidered magnifying glass and yeoman's quill, with the quill nib pointing down and forward.

Operations Technician - Gold embroidered ship's helm, circumscribing an arrow which points diagonally up and forward. One spark passes diagonally downward through the helm (right and left).

Ordnance Technician - Gold embroidered, flaming, spherical shell, flame upward, perpendicular to the stripe.

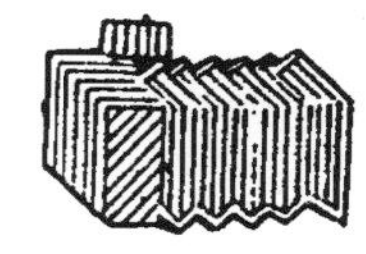

Photographer - Gold embroidered camera, with bellows extended, in an upright position with camera front toward sleeve front.

Security Technician - Gold embroidered star pointing up, in a circle, within a shield.

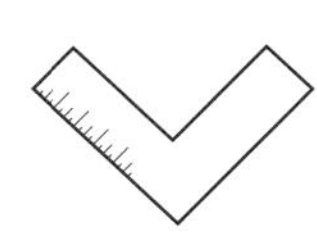

Repair Technician - Gold embroidered carpenter's square, point down, with arm inscribed with measurement lines to the front (right and left).

Ship's Clerk - Two gold embroidered, crossed quills, with points down and parallel to stripe.

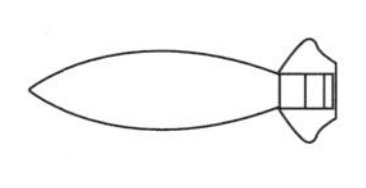

Underwater Ordnance Technician - Gold embroidered torpedo, with the torpedo parallel to the stripe, and warhead to the front (right and left).

Civil Engineer Corps - Same design and placement as for commissioned officers insignia.

Physician's Assistant and Technical Nurse Warrant Officer - Gold embroidered caduceus, with staff perpendicular to the stripe.

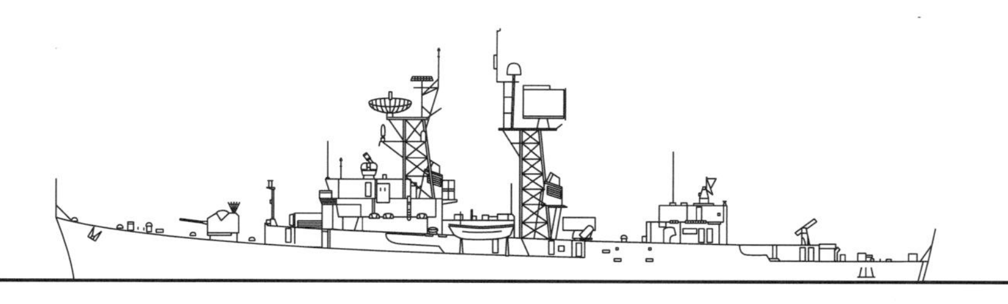

Enlisted Rank/Rate Insignia

The use of the word "rank" for Navy enlisted personnel is incorrect. The term is "rate." The rating badge is a combination of rate (pay grade) and rating (specialty) and is worn on the left sleeve of all uniforms in grades E-4 through E-6. Group rate marks for E-1 (optional) through E-3 are worn on the dress uniforms only. Chief petty officers (E-7 through E-9) wear collar devices on their white and khaki uniforms, and rate badges on their service dress blues. The rating badges consist of a perched eagle (sometimes referred to as a "crow") with expanded wings pointing upward and the eagle's head to its left. Chevrons (rocker, and stars for CPOs), indicating the wearer's rate, and a specialty mark to indicate rating are included as part of the rating badge. Group rate marks consists of two or three diagonal stripes, either alone or with specialty marks indicating E-2 and E-3 paygrades. Chief petty officers wear their rank devices in different places on their uniforms, depending upon the uniform. The three basic uniforms and the type of rank devices are: khakis (a working uniform) - pins on the collar; whites - pins on the collar; and, blues - stripes sewn on the upper sleeve. Soft shoulder boards are worn on white epauletted shirts when worn with Service Dress Blue, and on Navy blue/black pullover sweater (wooly-pulley). The collar device is also worn on the left side of the garrison cap.

Pay Grade	Rate	Abbreviation	Upper Sleeve	Collar and Cap
E-9	Master Chief Petty Officer of the Navy	MCPON		
E-9	Master Chief Petty Officer	MCPO		
E-8	Senior Chief Petty Officer	SCPO		
E-7	Chief Petty Officer	CPO		
E-6	Petty Officer First Class	PO1		
E-5	Petty Officer Second Class	PO2		
E-4	Petty Officer Third Class	PO3		
E-3	Seaman, Fireman, Constructionman, Airman, Hospitalman, Dentalman	SN, FN, CN, AN, HN, DN		none currently
E-2	Seaman Apprentice, Fireman Apprentice, Construction Apprentice, Airman Apprentice, Hospital Apprentice, Dental Apprentice	SA, FA, CA, AA, HA, DA		none currently
E-1	Seaman Recruit	SR		none currently

MASTER CHIEF PETTY OFFICER OF THE NAVY - Sleeve Insignia: Three gold stars above a perched eagle, rocker and chevrons. The Master Chief Petty Officer of the Navy has a gold star in place of the specialty mark. Post tour MCPON continue wearing three gold stars above the eagle. A gold specialty mark is worn in lieu of the gold star in the center of rating badge unless assigned to a Command Master Chief billet, then a silver star replaces the specialty mark. MCPO's assigned to a full time CM/C billet wear silver stars above the eagle and a silver star in place of the specialty mark. Post Tour CM/C's wear two silver stars above the eagle and a silver specialty mark in the center of the rating badge.

Metal Grade Insignia: Gold fouled anchor with silver block letters "USN" super-imposed on the shank of the anchor. Three silver stars are attached above the anchor stock to designate Master Chief Petty Officer of the Navy.

MASTER CHIEF PETTY OFFICER - Sleeve Insignia: Two silver stars above a perched eagle, rocker and chevrons. Fleet/Force Master Chief Petty Officers wear two gold stars above the eagle and one gold star in place of the specialty mark. Post Tour FM/C's continue wearing two gold stars above the eagle. A gold specialty mark replaces the gold star centered on the rating badge unless assigned to a Command Master Chief billet following, then a silver star replaces the specialty mark. Command Master Chief Petty Officer. Senior Chief Petty Officers and Chief Petty Officers filling CM/C billets do not replace the specialty mark with a star. Post Tour CM/C's wear two silver stars above the eagle and a silver specialty mark in the center of the rating badge (excluding past MCPON's and FM/C's).

Metal Grade Insignia: gold fouled anchor with silver block letters "USN" super-imposed on the shank of the anchor. Two silver stars are attached above the anchor stock to designate Master Chief Petty Officer.

SENIOR CHIEF PETTY OFFICER - Sleeve Insignia: One silver star above the eagle chevrons and rocker.

Metal Grade Insignia: Gold fouled anchor with silver block letters USN super-imposed on the shank of the anchor. One silver star is attached above the anchor stock to designate Senior Chief Petty Officer.

CHIEF PETTY OFFICER - Sleeve Insignia: Eagle chevrons and rocker.

Metal grade insignia: gold fouled anchor with silver block letters USN super-imposed on the shank of the anchor. Collar grade insignia are worn on both collar points on long sleeve khaki, blue shirt or coveralls. The insignia is centered 1 inch from the front and upper edges of the collar. This procedure applies whether the collar is worn open or closed. On the open collar short sleeve shirt, the insignia is centered at a point 1 inch from the front and lower edges of the collar and position the vertical axis of the insignia along an imaginary line bisecting the angle of the collar point. Men wear collar insignia on the standing collar of the service dress white coat. The anchor shank will be parallel to the vertical edge of the collar, with the center of the insignia on the midline of the standing collar, 1 inch from the vertical edge of the collar. Women wear collar insignia on the ends of the collar of the service dress white coat. The anchor shank is to be in the vertical (upright) position, with the center of the insignia approximately 1 inch from the bottom edge of the collar and midway between the edges (seam and outer edge) of the collar.

PETTY OFFICER - Sleeve insignia: Rating badges consist of a perched eagle with expanded wings pointing upward and its head facing right. Chevrons, indicating the wearer's rate, and a specialty mark indicating rating are part of the badge. Rating badges worn on blue working jackets and chambray shirts have no specialty mark. Iron-on, sew-on, or embroidered on rating badges are permitted on chambray shirts. Chevrons on rating badges for men, E-4 through E-6, measure 3-1/4 inches wide. Chevrons on women's rating badges measure 2-1/2 inches wide and their rating insignia is 3/4 the size of men's. Women wearing the men's peacoat will use the men's size rating badge on the peacoat. Men and women wear the same size rating badge (3-1/4 inches wide) on the blue working jacket.

E-2 AND E-3 PAYGRADES (GROUP RATE MARKS) - Sleeve Insignia: Badge consists of two or three short diagonal stripes which, alone, or in combination with specialty marks, indicate E-2 and E-3 paygrades. Personnel in paygrade E-1 do not wear group rate mark. Men and women wear the same size.

Group rate marks are placed on a rectangular background and worn on the left sleeve of all uniforms except coveralls and dungarees. They are worn in the same relative position as that of rating badges. Group rate marks are not worn on any outer garment. The stripes are 3 inches long and placed at an angle of 30 degrees from the horizontal line on a rectangular background of a color that matches the uniform on which it is worn. The lower end of the stripes is to the front. E-2 personnel wear two stripes and E-3 personnel wear three stripes.

Seamen and seamen apprentices wear white stripes on blue uniforms and navy blue stripes on white uniforms. Firemen and firemen apprentices wear red stripes on blue and white uniforms. Constructionmen and construction apprentices wear light blue stripes on blue and white uniforms. Airmen and airmen apprentices wear emerald green stripes on blue and white uniforms. Hospitalmen, dentalmen and apprentices wear white stripes and specialty marks on blue uniforms and navy blue stripes and specialty marks on white uniforms.

Striker's Mark (Gunner's Mate)

STRIKERS - Sleeve Insignia: E-1, E-2 and E-3 personnel who are qualified, and have been designated, wear the specialty mark (see page 23) of the rating for which they have qualified. E-l's currently wear the striker mark only, centered 2 inches above the midway point between shoulder and elbow on the left sleeve of all uniforms except dungarees, in the same position relative to the center line of the sleeve as prescribed for rating badges. E-2's and E-3's wear the striker mark centered immediately above the background of the group rate marks on the left sleeve, or striker mark and group rate mark may be one piece. They are worn on all uniforms except dungarees and outer garments. Men and women wear the same size.

APPRENTICE TRAINING GRADUATES - Sleeve Insignia: A winged circle with an anchor in the center (Airman), a machine valve wheel with a wrench (Fireman), and an anchor (Seaman). These marks are worn by non-rated graduates of formal training, whose designation as a striker has not yet been authorized. Apprentice training graduates wear the devices in the same relative position as the striker marks. Apprentice training graduates wear the devices until designated a striker then replace apprentice devices with an appropriate striker mark (Airmen Firemen, Seamen). *These marks are also shown with specialty marks on page 23.*

UNIT IDENTIFICATION MARKS (UIM's) - Sleeve Insignia: E-1 through E-6 personnel assigned for permanent duty (including Naval Reserve Reinforcement and Augment Personnel) are required to wear UIM's on the right sleeve of dress jumper uniforms, winter blue shirts, and short sleeved white shirts. UIM's have 1/4 inch white block letters, embroidered on a black background 1/2 inch wide, and are worn with the top edge parallel to and 3/8 inch below lower row of shoulder sleeve stitching. UIM's are centered on the outer face of the sleeve and sewn with colorfast blue thread. UIMs are authorized in two lengths, 5 inch and $5^3/_4$ inch. UIMs are not required when in transit between duty stations.

Service Stripes

Enlisted service stripes consist of embroidered diagonal stripes, 7 inches long and $^3/_8$ inch wide for male CPO's. Male E-l through E-6 personnel wear service stripes $5^1/_4$ inches long and $^3/_8$ inch wide. Navy women wear service stripes $5^1/_4$ inches long and $^1/_4$ inch wide.

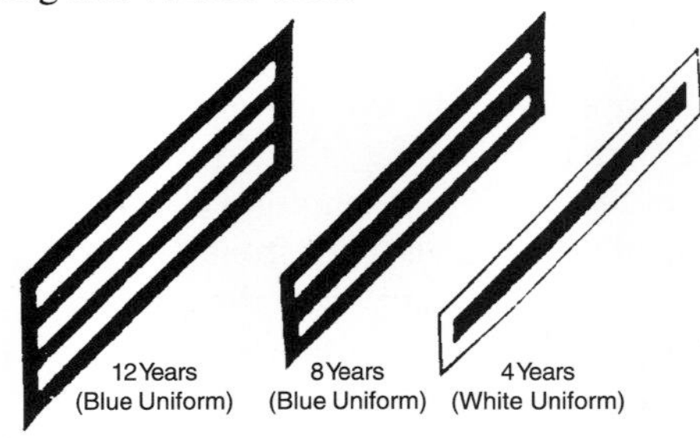

The service stripe are positioned on the left sleeve of dress blue, dress white, dinner dress blue jacket, and the dinner dress white jacket with the lower ends to the front. The lower end of the first stripe is 2 inches from the end of the sleeve. On jumpers having a buttoned cuff, the lower end of the first stripe is 1-1/2 inches above the upper edge of the cuff. The trailing edge of the stripe is in line with the trailing edge of the rating badge. The stripes are at a 45 degree angle. When more than one stripe is authorized they are placed 1/4 inch apart.

Stripes are either scarlet, gold or blue. All enlisted personnel wear one stripe for each four years of active service (regular or reserve) in any of the armed services. Personnel whose most recent 12 cumulative years of Naval active or active reserve service meets requirements for Good Conduct Service (that which meets minimum requirements for performance, conduct and evaluations marks for the Good Conduct Medal) shall wear gold rating badges and gold service stripes on dress blue uniforms and dinner dress blue uniforms and dinner dress blue/white jacket uniforms. The 12 years may be active or drilling reserve time in the Navy, Navy Reserve, Marine Corps, or Marine Corps Reserve. Times excluded are: delayed entry programs, inactive reserves and broken service. Under broken service conditions - service resumes with the cumulative time count upon active duty reenlistment, or upon enlistment in the drilling reserves.

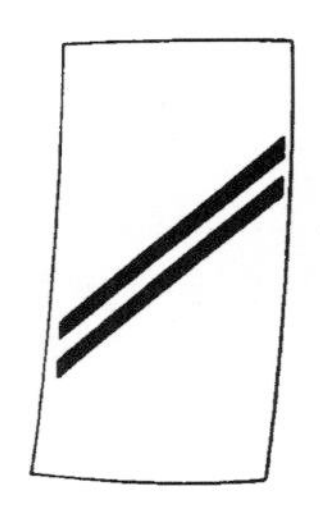

Coats

Jumpers

Specialty Marks

Specialty marks have been used in the United States Navy since 1886. The mark is the central element of the rating badge, which is a combination of rate and rating. The badge indicates an individual's occupational specialty, or rating (the chevrons on the badge indicate the individuals rate). Current rating badges and specialty marks are blue on the white uniforms and white on the blue uniforms, with blue chevrons on the white uniforms and scarlet chevrons on the blue uniforms. Current specialty marks, indicating the rating, are centered between the eagle's talons and the upper chevrons, or as a designator (striker mark) above group rate marks indicating that the individual is a striker. The specialty marks shown here cover the period from World War II to present.

Some specialty marks were also used as distinguishing marks for non-rated men qualified for the rating (strikers) and are so noted.

Aerographer's Mate (Aerographer from 1923 to 1942)- A winged circle with vertical feathered arrow through it. One-half of the circle is filled in and to the front. *This mark was also used as a distinguishing mark.*

Air Traffic Controller (Air Traffic Controlman) - A winged microphone. (Established in 1948 from the ratings of Specialist (Y) Control Tower Operators, Radarman, Specialist (X) Operations-Plotting and Chart Work and Specialist (V) Transport Airman.)

Aircrew Survival Equipmentman - A winged parachute. (Title established in 1942 as parachute rigger; title changed to Aircrew Survival Equipmentman in 1965.)

Airship Rigger - A winged anchor with and air ship. (Rating established in 1943; disestablished in 1948.)

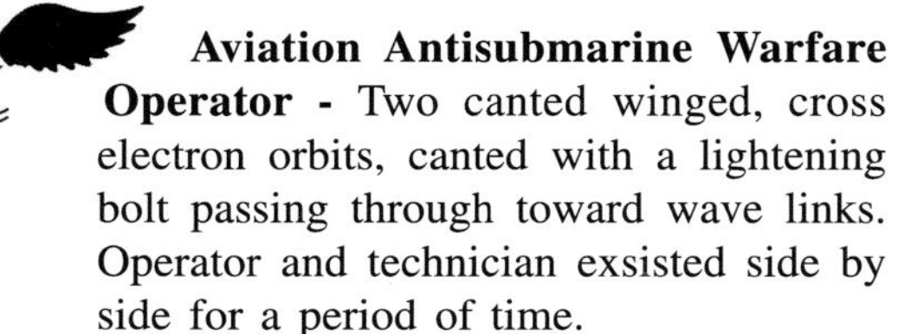

Aviation Antisubmarine Warfare Operator - Two canted winged, cross electron orbits, canted with a lightening bolt passing through toward wave links. Operator and technician exsisted side by side for a period of time.

Aviation Antisubmarine Warfare Technician - A lightning bolt pointing toward winged water and an arrow below water surface pointing downward. (Rating established in 1963; disestablished in 1990 and converted to Aviation Electronics Technician.)

Aviation Boatwain's Mate - Crossed winged anchors, crowns down. (Rating established in 1944.)

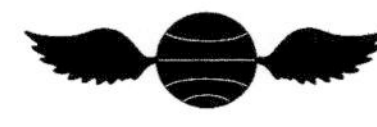

Aviation Electrician's Mate - A winged globe, with five embroidered latitudinal lines and five embroidered longitudinal lines. (Rating established in 1942.)

Aviation Electronicsman - Winged sparks; points to the front. (Rating established in 1948 from Aviation Radioman to Aviation Electronics Technician in 1959. Rating was disestablished in 1959)

Aviation Electronics Technician - Winged helium atom, surrounded by revolving electrons, one horizontal and one vertical. (Rating established in 1948.)

Aviation Fire Control Technician - A winged range finder. (Rating established in 1954; disestablished in 1990 and converted to Aviation Electronics Technician.)

Aviation Guided Missilman - A winged guided missile. (Rating established in 1954; disestablished in 1960.)

Aviation Machinist's Mate - A winged two-bladed propeller. (Rating established in 1921.) *This mark was also used as a distinguishing mark.*

Aviation Maintenance Administrationman - A winged two-bladed propeller centered on an open book. (Rating established in 1963.)

Aviation Ordnanceman - A winged flaming spherical shell. (Rating established in 1926.)

Aviation Photographer's Mate - Winged graphic solution of a photographic problem. (Rating established in 1948; combined with Photographer's Mate in 1953.)

Aviation Pilot - A representation of the Naval Aviator's wings. (Rating established in 1924; disestablished in 1935.) *This mark was also used as a distinguishing mark.*

Aviation Radioman - Winged sparks; points forward. (Rating included in Aviation Electronics Technician and Aviation Electronicsman in 1948.)

Aviation Radio Technician - Winged sparks; points forward. (Rating established in 1942, changed to Radio Technician's Mate in 1945.)

Aviation Storekeeper - Winged crossed keys; stems of keys down, webs facing outward. (Rating established in 1945 as Storekeeper (V); title changed in 1948.)

Aviation Structural Mechanic (Aviation Metalsmith) - Winged crossed mauls; heads of mauls up. (Rating first established as Aviation Metalsmith in 1921; title changed in 1948.) *This mark was also used as a distinguishing mark.*

Aviation Support Equipment Technician - A winged maul and spark, head of maul up; spark points down and to the front. (Rating established in 1966.)

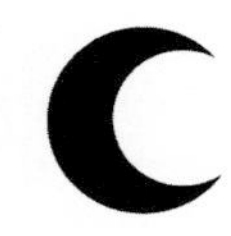

Baker - A crescent. (Rating established in 1908, combined with Ship's Cook, as Commissaryman in 1948.)

Blacksmith - Crossed mauls; heads of mauls up. (Rating disestablished in 1936.)

Boatswain's Mate - Crossed anchors; crowns down. (Rating established in 1885.)

Boilermaker - Hero's boiler discharge vents fitted so that cylinder would revolve in a clock-wise direction. (Rating re-established in 1956 and disestablished in 1977.)

Boilerman - Hero's boiler discharge vents fitted so that cylinder would revolve in a clock-wise direction. (Rating established in 1948, changed to Boiler Technician in 1976.)

Boiler Technician - Hero's boiler discharge vents fitted so that cylinder would revolve in a clock-wise direction. (Rating established in 1976; disestablished in 1996 and converted to Machinist's Mate.)

Bugler/Buglemaster - A bugle; bell to the left. (Rating established in 1927; included in Quartermaster in 1948.) *This mark was also used as a distinguishing mark.*

Builder - Carpenter's square, points up, superimposed on plumb bob. (Rating established from Carpenter's Mate (Builder) rating in 1948.)

Carpenter's Mate - Crossed axes; ax heads up. (Rating established in 1885; became Builder and Damage Controlman ratings in 1948.) *This mark was also used as a distinguishing mark.*

Commissaryman - Crossed keys over a quill pen. (Rating established in 1948 from Chief Commissary Steward, Ship's Cook, Ship's Cook (Butchers) and (Bakers); merged into Mess Management Specialist in 1974.)

Construction Electrician - A spark super-imposed, at an angle, on a telephone pole; lower end of spark to the front. (Rating changed from Construction Electrician's Mate in 1958.)

Construction Mechanic - Double-headed wrench superimposed on a nut. (Rating changed from Mechanic in 1958.)

Cook - A crescent. (Rating disestablished.) See Ship's Cook.

Coppersmith - Crossed mauls; heads of mauls up. (Rating disestablished.)

Coxswain - Crossed anchors; crowns down. (Rating included in Boatswain's Mate in 1948.)

Cryptologic Technician - Crossed quill and spark, both pointing down; pen on top with nib to the front. (Rating established in 1948 as Communications Technician; changed in 1976.)

Damage Controlman - A crossed axe and maul. (Rating established in 1948 from Specialist (F) Firefighter, Carpenter's Mate and Painter; disestablished in 1984.)

Data Processing Technician - Quill superimposed diagonally on gear; pen nib down and to the front. (Rating established in 1948 as Machine Accountant; changed in February 1967; disestablished in 1998 and converted to Information Technology Specialist.)

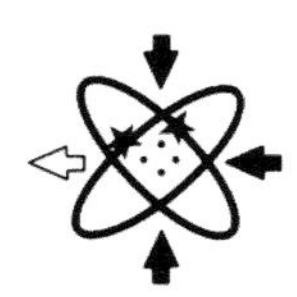

Data Systems Technician - A helium atom with three arrows pointing toward its center and one pointing out from its center. (Rating established in 1963; disestablished in 1998 and converted to Fire Controlman and Electronics Technician.)

Dental Technician - A caduceus, with a block letter "D" midway on the staff. (Rating established in 1948.)

Disbursing Clerk - A check with a key in it at an angle; web and pin of key down and to the front. (Rating established in 1948 from Storekeeper (D) Dispersing.)

Electrician's Mate - A globe, with five latitudinal and five longitudinal lines. (Rating established in 1921.) *This mark was also used as a distinguishing mark.*

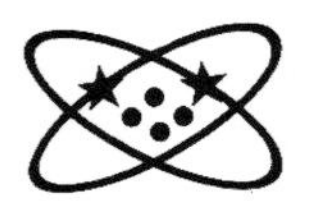

Electronics Technician - A helium atom. (Rating changed from Electronics Technician's Mate in 1948, which was changed from Radio Technician in 1945.)

Electronics Warfare Technician - A helium atom with an electrical spark passing downward through the atom. (Rating established in 1971.)

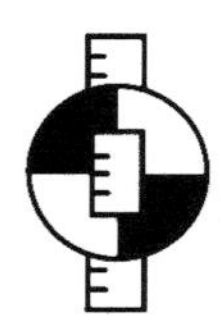

Engineering Aid - A leveling rod with the measuring scale to the front. (Rating changed from Surveyer in 1961.)

Engineman - A gear. (Rating established from Motor Machinist's Mate in 1948.)

Equipment Operator - A bulldozer, blade to the front. (Title changed from Driver in 1948.)

Fire Control Technician (Current) - A range finder. (Rating established from Fire Controlman in 1955.) *This mark was also used as a distinguishing mark those other than for strikers for Gun Rangefinder Operators and Seaman Fire Controlmen.*

Fire Control Technician (Obsolete) - A director with radar screen. This mark was abolished when the rating was combined with Fire Controlman under the title of Fire Control Technician in 1955.

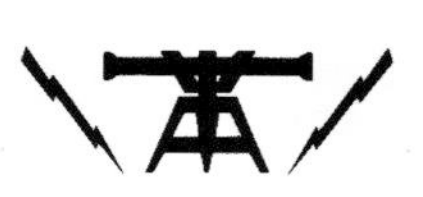

Fire Controlman - A range finder with spark on each side that faces inward. (Rating approved in 1983, effective 1985.)

Gas Turbine System Technician - A turbine with an impeller. (Rating approved in 1974 and held in abeyance until 1978.)

Gunner's Mate - Crossed gun barrels; muzzles up. (Rating established in 1885.) *This mark was also used as a distinguishing mark.*

Hospital Corpsman - A caduceus. (Rating changed from Pharmacist's Mate in 1948, which used the Geneva Cross as a mark.)

Hull Maintenance Technician - Crossed fire axe and maul, handles down; fire axe blade to front, on a carpenter's square that point down. (Rating established in 1972.)

Illustrator Draftsman - A triangle with draftsman's compass on it; the right angle of triangle points down. (Rating established as Draftsman in 1948. Title changed in 1961.)

Instrumentman - A pair of calipers with an adjusting screw. (Rating established in 1944 as Special Artificer and title changed in 1948; disestablished in 1999 and converted to Interior Communications Electrician, Electrician's Mate, Electronics Technician and Aviation Electronics Technician

Intelligence Specialist - Magnifying glass and quill, nib of pen down. (Rating established in 1975.)

Interior Communications Electrician - Electrician Mate's device with a French-style telephone above it. (Rating established in 1948 from Electrician's Mate ratings.)

Journalist - Scroll and quill; pen uppermost, nib of pen down and to the front. (Rating established in 1948.)

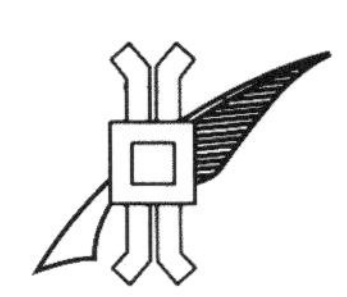

Legalman - A vertical mill rinde over a quill; nib of pen down and to the left. (Rating established in 1973.)

Lithographer - Crossed litho crayon holder and scraper uppermost, blade to the front. (Rating established in 1948.)

Machinery Repairman - Micrometer and gear; handle of micrometer to the rear, open parts of jaws holding gear. The device is worn with the handle parallel to the upper edge of the left arm of the chevron. (Rating established in 1948.)

Machinist's Mate - Three-bladed propeller; one blade pointing down. (Rating established as Machinist in 1885; title changed in 1904.) *This mark was also used as a distinguishing mark.*

Mailman - A postal cancellation mark with the letter "M." (Rating established to replace Specialist (M) MailClerk into Teleman in 1948.)

Master-at-Arms - A star pointing up in a circle, within a shield. (Rating established in 1974.)

Mess Management Specialist - Crossed keys and quill superimposed upon an open ledger. (Rating established in 1975.)

Metalsmith - Crossed mauls; heads up. (Rating disestablished in 1959.)

Mineman - A floating mine. (Rating established in 1944.)

Missile Technician - A guided missile surrounded by an electronic wave. (Rating established in 1954; title changed from Guided Missileman in 1961.)

Molder - Crossed bench rammer and stove tool, pointing down; bench rammer uppermost; rounded end of rammer to the front. (Rating established in 1948, disestablished in 1997 and converted to Machinery Repairman.)

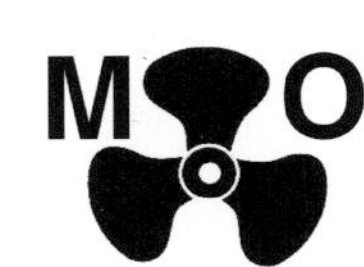

Motor Machinist's Mate - Three-bladed propeller; one blade pointing up, with the initials "M" and "O." (Rating established in 1942, combined into Engineman in 1948.)

Musician - A lyre. (Rating established in 1944 from Bandmaster and 1st Muscian.) *This mark was also used as a distinguishing mark.*

Navy Counselor - An anchor crossed with a quill. (Rating established in 1974.)

Nuclear Weaponsman - A bomb superimposed on an atom. (Rating Established in 1957; disestablished in 1961.)

Ocean Systems Technician - Neptune's trident rising through the waves. (Rating established in 1970; disestablished in 1997 and converted to Sonar Technician - Surface.)

Operations Specialist - A scope on an arrow; arrow pointing diagonally upward and to the front. (Rating established from Radarman in 1972.)

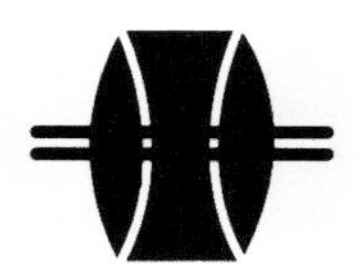

Opticalman - Double concave lenses between two double convex lenses; two lines passing horizontally through lenses symbolizing passing of light. (Rating established in 1948; disestablished in 1999 and converted to Interior Communications Electrician and Electrician's Mate.)

Painter - Crossed axes; axe heads up. (Rating established in 1885; included in Damage Controlman in 1948.)

Parachute Rigger - A winged parachute. (Rating established in 1942; changed to Aircrew Survival Equipmentman in 1965.)

Patternmaker - A wooden jack plane, facing front. (Rating established in 1948 from ratings wearing the Carpenter's Mate mark; disestablished in 1997 and converted to Hull Maintenance Technician.)

Personnelman - Crossed manual and quill; manual upper most; pen nib down and to the front. (Rating established in 1948.)

Pharmacist's Mate - A Geneva Cross. (Rating re-titled from Hospital Steward in 1916; changed to Hospital Corpsman with new mark in 1948.) *This mark was also used as a distinguishing mark for Hospital Apprentice.*

Photographer - A bellows camera. (Rating established in 1921; changed to Photographer's Mate in 1942 and mark changed to a graphic solution in 1948.)

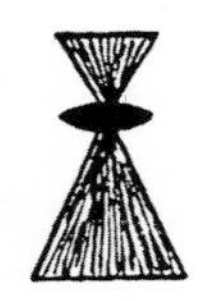

Photographer's Mate - A graphic solution of photographic problem. (Photographer's Mate established from Photographer in1942.) (Aviation Photographer's Mate combined with Photographer's Mate in 1953.) The current mark is a winged graphic solution; the rating without wings could be worn until Nov 1995.)

Photographic Intelligenceman - Magnifying device over a graphic solution. (Rating established in 1957; disestablished in 1975.)

Pipefitter - Crossed wrenches. (Rating established in 1948 merged into Shipfitter in 1958.)

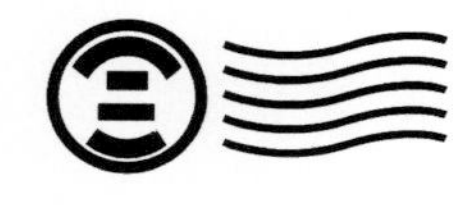

Postal Clerk - A postal cancellation mark. (Rating re-established from Teleman in 1960.)

Printer - An open book. (Rating included in Lithographer in 1955.)

Quartermaster - A ship's helm. (Rating established in 1885; Buglemaster included in the rating in 1948.) *This mark was also used as a distinguishing mark.*

Radarman - Three sparks superimposed on an arrow. (Rating established in 1942, included in Operations Specialist in 1972.)

Radio Technician - Four sparks; points to the front. (Rating established in 1942; included in Electronics Technician's Mate in 1945 and to Electronics Technician in 1948.)

Radioman - Four sparks; points to the front. (Rating established in 1921; changed in 1999 to Information Technology Specialist.) T*his mark was also used as a distinguishing mark for Seaman Radioman*

Religious Program Specialist - A rose compass, a globe, and an anchor. (Rating established in 1979.)

Shipfitter - Crossed mauls; heads up. (Rating established in 1902, combined with Damage Controlman to form Hull Maintenance Technician in 1972.) This mark was also used for Molder and Metalsmith.

Ship's Cook - A crescent. (Rating established in 1908; became Commissaryman in 1948.) Beginning in 1924, Officer's Stewards and Cooks wore the crescent mark above horizontal bars designating grades. T*his mark was also used as a distinguishing mark.*

Ship's Serviceman - Crossed key and quill; stem of key and pen nib down; pen to be upper-most; web and pin of key to the front. (Rating established in 1944.)

Signalman - Two upright crossed semaphore flags. (Rating established in 1921, merged into Quartermaster in 1948 and re-established in 1957.) T*his mark was also used as a distinguishing mark for Seaman Signalman.*

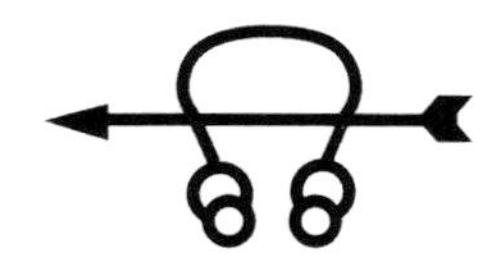

Sonar Technician - Earphones with arrow in horizontal position, point to the front. (Rating title changed from Sonarman in 1964. Originally established in 1942 as Soundman and changed to Sonarman in 1944.) *This mark was also used as a distinguishing mark, for Sonar Operator*

Specialists

The advent of World War II made it clear to the Navy that the peace-time General Service Rating structure did not provide the many skills needed to fight a global war, with rapidly expanding technical requirements. The Specialist Ratings were authorized as requirements developed, and skilled civilians were recruited and drafted to fill these billets. Many Specialist Ratings were converted to General Service Ratings at the end of the war in the 1948 reorganization of the rating system. Others were retained as Emergency Service Ratings held by men on inactive reserve and subject to recall for national emergencies. All these Specialist Ratings were disestablished by 1974.

Specialist (A) Physical Training Instructor, Airship Rigger, Aircraft Carburetor Mechanic - A diamond with the initial "A." (Rating established during WWII; disestablished.)

Specialist (B) Master At Arms, Stevedore - A diamond with the initial "B." (Rating established after WWII; disestablished.)

Specialist (C) Classification Interviewer, Chaplain's Assistant - A diamond with the initial "C." (Rating established during WWII; disestablished.)

Specialist (E) Recreation Assistant, Motion Picture Service (Booker), Physical Training Instructor - A diamond with the initial "E." (Rating established after WWII; disestablished.)

Specialist (F) Fire Fighter - A diamond with the initial "F." (Rating established during WWII; disestablished.)

Specialist (G) Aviation Free, Antiaircraft and Gunnery Instructors - A diamond with the initial "G." (Rating established during WWII; disestablished.)

Specialist (H) Harbor Defense Sonarman - A diamond with the initial "H." (Rating established in 1959; disestablished 1965.)

Specialist (I) IBM Operator, Punched Card Account Machine Operator, Instructor (misc.) - A diamond with the initial "I." (Rating established during WWII; disestablished.)

Specialist (K) Chemical Warfareman, Telecommunications Censorship Technician, Information Security Specialist - A diamond with the initial "K." (Rating established after WWII; disestablished.)

Specialist (M) Mail Clerk, Underwater Mechanic - A diamond with the initial "M." (Rating established during WWII; disestablished.)

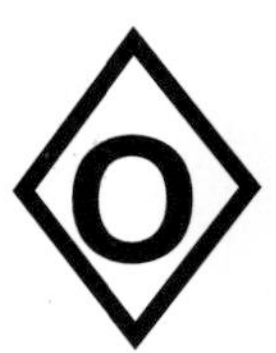

Specialist (O) Inspector of Naval Material - A diamond with the initial "O." (Rating established during WWII; disestablished.)

Specialist (P) Photographer, Photographic Specialist, Photogammetry Assistant - A diamond with the initial "P." (Rating established during WWII; disestablished.)

Specialist (Q) Communication Security, Communications Specialist, Cryptographer - A diamond with the initial "Q." (Rating established during WWII; disestablished.)

Specialist (R) Recruiter, Transportationman - A diamond with the initial "R." (Rating established during WWII; disestablished.)

Specialist (S) Entertainer, Shore Patrol and Security, Master at Arms (WAVE), Personnel Supervisor (V-10) - A diamond with the initial "S." (Rating established during WWII; disestablished.)

Specialist (T) Teacher, Instructor, Transportation Airman - A diamond with the initial "T." (Rating established during WWII; disestablished.)

Specialist (U) Utility(V-10) - A diamond with the initial "U." (Rating established during WWII; disestablished.)

Specialist (V) Transport Airman, Aviation Pilot - A diamond with the initial "V." (Rating established during WWII; disestablished.)

Specialist (W) Welfare (Chaplain's Assistant), Welfare and Recreation Leader - A diamond with the initial "W." (Rating established during WWII; disestablished.)

Specialist (X) Specialist Not Elsewhere Classified (49 different titles have been identified by this mark) - A diamond with the initial "X." (Rating established during WWII; disestablished.)

Specialist (Y) Control Tower Operator - A diamond with the initial "Y." (Rating established during WWII; disestablished.)

Steelworker - An I-beam suspended from a hook; open side of hook to the front. (Rating established in 1948.)

Steward - An open book with a key and wheat spike. This mark was established in 1963 as a replacement for the crescent mark previously used for Stewards, Ship's Cooks and Bakers (see Ship's Cook). (Rating merged with Commisaryman into Mess Management Specialist.) In 1924, Officer's Stewards and Cooks wore the crescent mark above horizontal bars designating grades.

Steward - A Crescent. (Rating established as Officer's Steward in 1924, changed to Steward in 1944, which was given the above mark in 1963. The rating was merged into Mess Management Specialist in 1975.)

Storekeeper - Crossed keys, stems down, webs outward. (Rating established in 1916.) *This mark was also used as a distinguishing mark..*

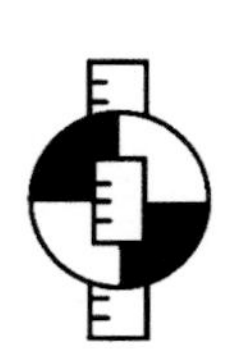

Surveyor - A leveling rod with the measuring scale to the front. (Rating established in 1948; changed to Engineering Aid in 1961.)

Teleman - Crossed quill pen and lightening bolt over a postal cancellation mark. Rating established in 1948; merged into Yeoman and Radioman in 1956; disestablished.)

Torpedoman's Mate - A torpedo, head to the front. (Rating established in 1921 as Torpedoman; changed to Torpedoman's Mate in 1942.) *This mark was also used as a distinguishing mark.*

Tradevman (Training Devices Man) - A lightening bolt passing through a gear. (Rating established in 1948; disestablished in 1988.)

Turret Captain - A profile of turret; barrel to the right. (Rating established in 1903; absorbed into Gunner's Mate in 1948.)

Underwater Mechanic - A diver's helmet over a two headed wrench. (Rating established and disestablished in 1948, but probably never activated.)

Weapons Technician - A spark and exploding shell on a trident. (Rating established in 1984; disestablished in 1996 and converted to Gunner's Mate, Mineman and Torpedoman's Mate.)

Utilities Man - A valve with flange to the front. (Rating established in 1948.)

Yeoman - Crossed quills, nibs down. (Rating established in 1896.) *This mark was also used as a distinguishing mark.*

Distinguishing Marks (Obsolete)

The term "distinguishing marks" was introduced into uniform regulations in 1905 (prior to that time they were referred to simply as "marks"). In 1922 uniform regulations stated that distinguishing marks were sleeve markings for men who had met certain qualifications, In addition to those required for their rating, or members of a crew that had obtained a special merit in competition. The marks were cloth insignia; white on blue for blues and blue on white for whites.

Uniform regulations of 1941 stated that several specialty marks could also be worn as distinguishing marks by men who had successfully completed a full course of instruction, or who had passed an examination for petty officer 3rd class, but had not yet selected for promotion. These individuals were called strikers for that rating. In 1941 uniform regulations stipulated that these "striker marks" would be worn midway between the wrist and the elbow on the right sleeve of for men of the seaman branch and on the left sleeve for others. In July of 1944, the Bureau of Personnel changed the position for wearing distinguishing marks (as specialty marks for non-rated men) to midway between the shoulder and the elbow. The use of specialty marks as distinguishing marks was completely changed with the introduction of group rates in 1948. The group rates included striker marks for qualified non-rated personnel, worn above the diagonal stripes.

In 1948 distinguishing marks were worn on the right sleeve, midway between the shoulder and the elbow. Distinguishing marks were abolished with the uniform revision of 1 July 1975 and in some cases replaced a badge to recognize a special qualification.

Specialty marks which were used as distinguishing marks are noted in the Specialty Marks section of this book and are not repeated in this section.

Advanced Undersea Weaponsman - A mine superimposed on the center of a torpedo, with three sparks pointing upward from the top of the mine; head of the torpedo to the front.

Aircraft Gunner - A winged machine gun; muzzle up. The machine gun was originally slanted at an angle of 60 degrees, but was changed to vertical in 1946 to accommodate the Aircraft Machine Gunner, first class, which had a $^1/_4$ " star above the muzzle.

Aircrewman - A winged circle with the initials "AC" in the center.

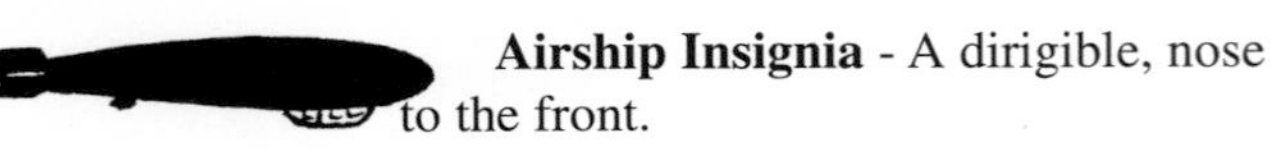

Airship Insignia - A dirigible, nose to the front.

Amphibious Insignia - A scarlet shoulder patch with a gold machine gun over an eagle on top of an anchor. Earlier unauthorized insignia included patches showing tanks being disgorged from an alligator's mouth. All shoulder patches were abolished in 1947.

Anti-Aircraft Machine Gunner - An approaching plane, centered in the cross wires of a gunsight; nose of the plane to the right.

Armed Guard - The initials "AG."

Assault Boat Coxswain - An arrow head centered on crossed anchors.

Aviation General Utility - Wings with a small "O" at the center. Originally designated a distinguishing mark for non-rated men assigned to aviation units. It was also listed as a specialty mark in 1941 Uniform Regulations to be worn by non-rated graduates of aviation general utility courses.

Bombsite Mechanic- The initial "B" between bombsite marks.

Construction Battalion - The initials "CB." This mark was replaced in favor of the multi-colored shoulder patch approved for Seabees during World War II.

Construction Battalion - The "fighting bee" shoulder patch. This patch, with all shoulder patches adopted by the Navy, was abolished on 17 January 1947.

Diver 1st Class, 2nd Class and Salvage - A diver's helmet with a "1," "2," or "S" on the breast plate. Originally this insignia had no number or initial, but they were added in 1944 when Diver 2nd Class (with a number "2") and Salvage Diver marks were introduced. Diver 1st and 2nd class and Salvage marks were discontinued in favor of a metal qualification breast insignia in 1973. The helmet with the block initial "D" and Initials "SD" were also used to identify Deep Sea and Scuba Divers.

Master Diver - A diver's helmet with the initial "M" on the breast plate. This mark was discontinued in favor of a metal qualification breast insignia in 1969.

Ex-Apprentice mark - An loosely intertwined figure-8 knot.

Expert Lookout - Binoculars.

Explosive Ordnance Disposal Technician - A mine over a crossed torpedo and bomb.

Fire Control Radar Operator - A range finder with a radar screen superimposed.

Fire Fighter Assistant - A fireman's Maltese cross.

Gun Captain - A horizontal naval rifle (gun) barrel.

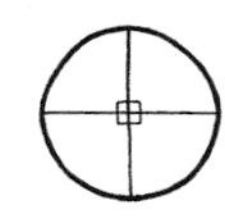

Gun Pointer, First Class and Gun pointer (without the star) - A gun sight and star.

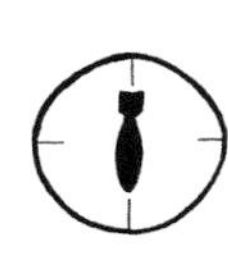

Master Horizontal Bomber - A star over a bomb site with a bomb pointing down.

Mine Assemblyman - A mine and anchor block. Originally established as Mine Warfare Insignia; the title was changed in 1958.

Minecraft Personnel Shoulder Patch - A multi-color shoulder patch with a mine on a wave with three lightening bolts.

Navy E - An block initial "E." Subsequent awards have a horizontal bars below the "E." Multiple award bar were changed from horizontal to diagonal in 1957. In 1960, a gold "E" and multiple award bars were authorized for the fifth and subsequent awards

Officer's Stewards and Cooks - A crescent. Horizontal bars were added to indicate grade (3rd class - one bar, 2nd class - two bars, 1st class - three bars and chief - four bars).

Ordnance Battalion - The initials "OB."

Parachuteman - An open parachute. This mark appeared in the 1941 Uniform Regulations and was worn prior to the establishment of the Parachute Rigger rating. It should be noted that there is no record either originating or abolishing this mark.

Patrol Torpedo Boat - The initials "PT." This mark was discontinued in favor of a shoulder patch with a white torpedo aimed at a downward angle and a rope border on a blue background

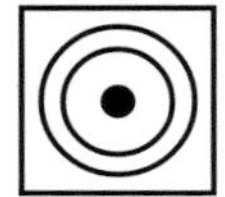

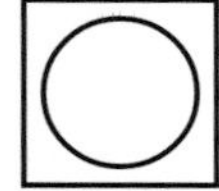

Rifle and Pistol Qualification Marks - An "Able" target (Expert - two rings and a bull's-eye, Sharpshooter - one ring and a bull's-eye and Marksman - one ring).

Seaman Gunner - A bursting bomb.

Submarine Insignia - Bow quarter of a submarine between two dolphins. This mark was replaced with the metal breast insignia in 1950.

Bit of History

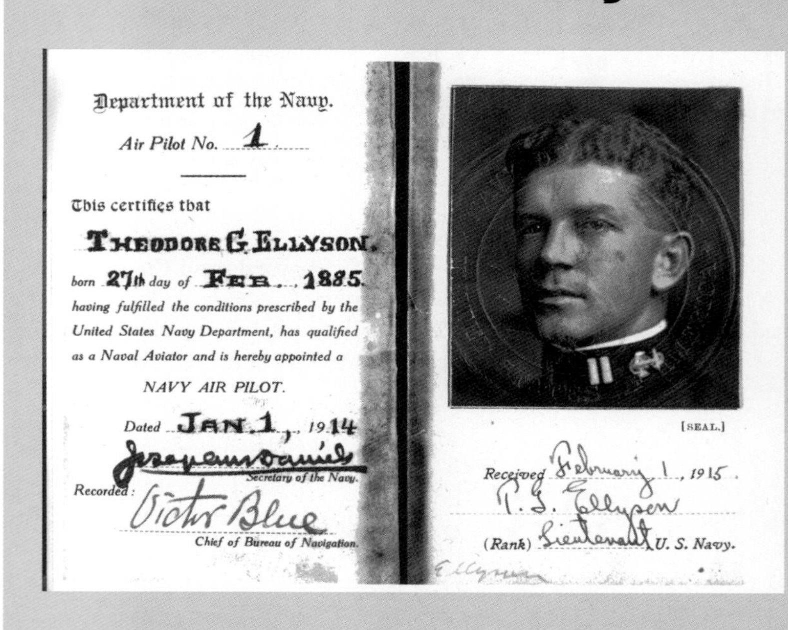

Department of the Navy.

Air Pilot No. 1

This certifies that

THEODORE G. ELLYSON.

born 27th day of FEB. 1885.

having fulfilled the conditions prescribed by the United States Navy Department, has qualified as a Naval Aviator and is hereby appointed a

NAVY AIR PILOT.

Dated JAN. 1, 1914

Josephus Daniels
Secretary of the Navy.

Recorded: Victor Blue
Chief of Bureau of Navigation.

[SEAL.]

Received February 1, 1915.

T. G. Ellyson

(Rank) Lieutenant U. S. Navy.

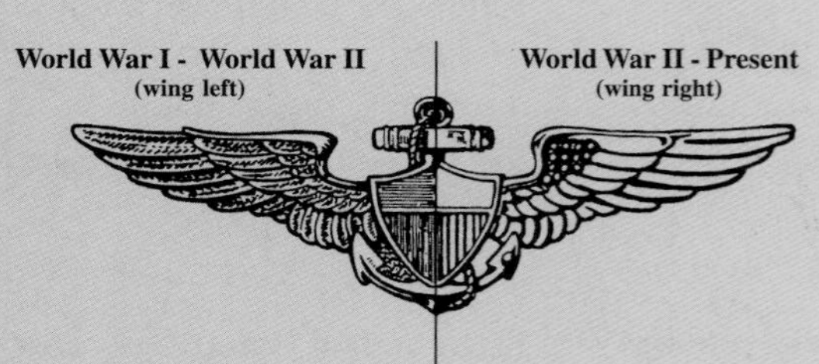

The fact that the Navy's oldest qualification badge has changed little in over a hundred years reflects the emphasis the Navy places on tradition.

Breast insignia are worn by Naval personnel who are qualified/designated in selected areas. Breast insignia are worn on the left breast of all service and dress coats, jumpers and khaki shirts. Miniature insignia, one-half regular size, are worn on formal and dinner dress jackets.

Naval Astronaut Insignia - A gold embroidered or gold metal winged pin with a star shooting diagonally from the bottom right to the top left, through an elliptical ring, on the shield of traditional Naval Aviator's wings.

Naval Astronaut (NFO) Insignia - A gold embroidered or gold metal winged pin with a star shooting diagonally from bottom right to top left through an elliptical ring, on the shield of the traditional Naval Flight Officer's wings.

Naval Aviator Insignia - A gold embroidered or gold metal winged pin with a fouled anchor behind a shield in the center.

Balloon Pilot Insignia (Obsolete) - A gold embroidered or gold metal half-winged pin with a fouled anchor behind a shield in the center.

Naval Aviation Observer and Flight Meteorologist Insignia - A gold embroidered or gold metal winged pin with a silver 0 circumscribing plain silver anchor on a gold background.

Naval Aviation Observer (Tactical) Insignia (Obsolete) - A gold embroidered or gold metal pin consisting of basic wings with a silver centerpiece superimposed upon two gold crossed fouled anchors. The silver centerpiece device has two crossed guns (naval rifles) superimposed upon it in bold relief and in gold color. This insignia became obsolete on 18 March 1947 and has been replaced by the current Naval Aviation Observer Insignia.

Naval Aviation Observer (Navigator) Insignia (Obsolete) - Gold embroidered or metal pin consisting of basic wings with a central device of two fouled anchors surmounted by a replica of a mariner's compass, superimposed on a silver-colored disk.

Naval Aviation Observer (Radar) Insignia (Obsolete) - A gold embroidered or gold metal pin consisting of basic wings with a silver centerpiece superimposed upon two gold crossed fouled anchors. The silver centerpiece device has a symbolic radar manifestation in gold bold relief superimposed upon it. This insignia became obsolete on 18 March 1947.

Naval Aviation Supply Corps Insignia - A gold embroidered or gold metal winged pin with a supply corps oak leaf in the center.

World War II

Naval Flight Surgeon Insignia - A winged, gold embroidered or gold metal pin with a Medical Corps device in the center inscribed in an oval. The World War II insignia was similar to the current pin, except that the wings were stylized.

World War II

Naval Flight Nurse Insignia - A gold metal pin designed like the flight surgeon's without the acorn. The World War II insignia had stylized wings, similar to the World War II Flight Surgeon Insignia. The World War II insignia was similar to the current pin, except that the wings were stylized.

Naval Flight Officer Insignia (NFO) Insignia - A gold embroidered or gold metal winged pin with a shield with a set of small, crossed, fouled anchors in the center.

Naval Aviation Experimental Psychologist and Aviation Physiologist Insignia - A gold embroidered or gold metal winged pin with a Medical Service Corps device in the center inscribed in an oval.

Naval Aircrew Insignia - A gold metal winged pin with an anchor inscribed in a circle in the center. The letters "AC" straddle the shank of the anchor.

Combat Aircrew Insignia (Obsolete) - An oxidized silver-colored winged metal pin, with a gold-colored circular shield with a superimposed fouled anchor; the word "AIRCREW" in raised letters on a silver-colored background below the circular shield; above the shield is a silver-colored scroll. Gold stars, up to a total of three, as merited, were mounted on the scroll, necessary holes being pierced to receive them. A silver star was used in lieu of three gold stars. A maximum of three combat stars were awarded for display on the Combat Aircrew Insignia. This insignia, although still authorized for Marines, became obsolete for the Navy in 1968.

Naval Aviation Warfare Specialist Insignia - A silver metal winged pin with an anchor on a shield in the center and a scroll at the bottom of the shield.

Submarine Insignia - A gold or silver embroidered or metal pin, showing bow view of a submarine proceeding on the surface with bow planes rigged for diving; flanked by dolphins in horizontal position, their heads resting on upper edge of bow planes. Gold for officers and silver for enlisted.

Submarine Medical Insignia - A gold color metal pin with two dolphins facing an oval in the center inscribed with the Medical Corps device. The acorn in center is silver.

1950-52

Submarine Engineering Duty Insignia - A gold metal pin with two dolphins facing a silver circle center inscribed with a silver three bladed propeller on a gold background. The tips of propeller blades trisect the circle and one blade is vertical. During the 1950-52 period, the insignia was the Submarine Insignia with a block letter E in the center.

Submarine Supply Corps Insignia - A gold metal pin with two dolphins facing a Supply Corps oak leaf in the center.

Submarine Combat Patrol Insignia - A silver metal pin showing, the broadside view of a "Flying Fish" class submarine proceeding on the surface, with a scroll at the bottom of the wave. Gold stars are mounted on the scroll to indicate each successful patrol subsequent to that for which the original insignia was awarded. Stars may also be placed on the wave area of the insignia. Holes are bored for that purpose. A silver star indicates five successful patrols.

SSBN Deterrent Patrol Insignia - A silver or gold metal pin, showing the broadside view of a "Lafayette" class SSBN proceeding submerged, with a Polaris missile circled by three electron paths, centered in the foreground, and a scroll across the bottom of missile and submarine. Gold or silver stars, with silver given priority, are mounted on the scroll in the order shown on diagram at left to indicate each additional patrol, as follows:

(1) Silver pin alone, without stars, one patrol.
(2) A gold star is added for each additional patrol (maximum four gold stars).
(3) Patrol pin with silver star, six patrols.
(4) A gold star is added for each additional patrol (maximum four stars).
(5) Two silver stars are added for eleven patrols.
(6) A gold star is added for each additional patrol (maximum four gold stars).
(7) Pin with three silver stars, sixteen patrols.
(8) A gold star is added for each additional patrol (maximum three gold stars).
(9) Gold patrol pin, 20 patrols.
(10) Sequence is repeated adding gold stars for each additional patrol (maximum four gold stars) and a silver star for each additional five patrols.

Surface Warfare Insignia - A gold metal pin, with the bow and superstructure of a modern naval warship on two crossed swords, on a background of ocean swells.

Enlisted Surface Warfare Inignia - A silver metal pin, with the bow and superstructure of a modern naval warship on two crossed cutlasses, on a background of ocean swells.

Surface Warfare Dental Corps Insignia - A gold metal pin, with a spread oak leaf, a silver acorn on each side of the stem on two crossed swords, on a background of ocean swells.

Surface Warfare Medical Corps Insignia - A gold metal pin, with a spread oak leaf surcharged with a silver acorn on two crossed swords, on a background of ocean swells.

Surface Warfare Medical Service Corps Insignia - A gold metal pin, with a spread oak leaf, attached to a slanting twig on two crossed swords, on a background of ocean swells.

Surface Warfare Nurse Corps Insignia - A gold metal pin, with a spread oak leaf on two crossed swords, on a background of ocean swells.

Surface Warfare Supply Corps Insignia - A gold metal pin with a supply corps oak leaf centered on the bow and superstructure of a modern naval warship superimposed on two crossed naval swords, on a background of ocean swells.

Command-at-Sea Insignia - A gold metal pin consisting of a five-pointed pyramidal star on anchor flukes and a partially unfurled commission pennant showing six stars.

Command Ashore/Project Manager Insignia - A gold metal, three pronged trident centered on an elliptically shaped laurel wreath.

Small Craft Insignia - A gold or silver metal pin with a small craft circumscribed by an anchor flukes on the sides and bottom and a three star pennant on top.

Craftmaster Insignia - An enameled pin with ship's helm with two crossed fouled anchors in the center.

Special Operations Insignia - Same ship and bow wave as Surface Warfare Insignia. In lieu of crossed swords, there is an ordnance disposal bomb over crossed lightning rays on one side, and a diving helmet over two tridents on the other.

Special Warfare Insignia - A gold metal pin with an eagle holding a trident and a handgun, in front of an anchor. Originally this insignia was gold for officers and silver for enlisted, but was approved in gold for enlisted personnel in 1972.

Underwater Demolition Team Insignia (Obsolete) -A metal pin with trident and a handgun, in front of an anchor. Gold for officers and silver for enlisted. This insignia became obsolete in 1972.

Seabee Combat Warfare Specialist Insignia - A gold or silver metal pin with a bee on crossed sword and rifle superimposed on an anchor on a background of leaves.

Basic Parachutist Insignia - A silver metal pin with an open parachute flanked on either side by wings that curve upward.

Naval Parachutist Insignia - A gold embroidered or gold metal winged pin with a gold open parachute in the center.

Diving Officer Insignia - A gold metal pin with two upright seahorses facing a diving helmet, and two tridents projecting upward and canted outward form the diving helmet's cover. A double carrick bend superimposed on the breast plate.

Diving (Medical) Insignia - Same as the Diving Officer Insignia with a caduceus on the breast plate.

Master Diver Insignia - Same as the Diving Officer Insignia, but silver.

Diving Medical Technician Insignia - Same as Diving (Medical) Insignia, but silver in color.

First Class Diver Insignia - A silver metal diving helmet surrounded by sea serpents.

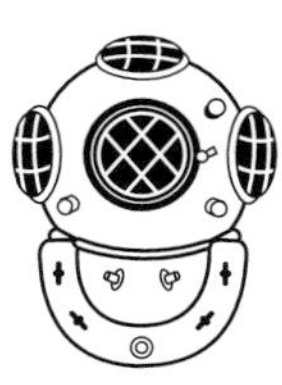

Second Class Diver Insignia - A silver metal diving helmet and breastplate.

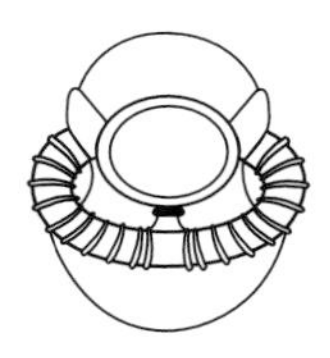

Scuba Diver Insignia - A silver metal pin wet suit hood and face mask with breathing apparatus.

Basic Explosive Ordnance Disposal Warfare Insignia - A silver metal pin with a conventional drop bomb, point down on a shield, and radiant with four lightning flashes, all within a wreath of laurel leaves.

Senior Explosive Ordnance Disposal Warfare Insignia - Same as the basic EOD Insignia, but with a star centered on the bomb.

Master Explosive Ordnance Disposal Warfare Insignia - Same as the Senior EOD Insignia, but with star in a laurel wreath affixed to the top of the shield.

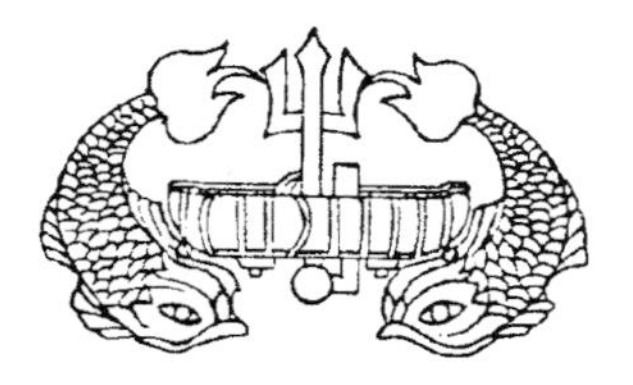

Deep Submergence Insignia - A gold or silver metal representation of the USS Trieste, on a trident, surrounded by dolphins.

Integrated Undersea Surveillance System Insignia - A gold or silver metal pin with a trident and seahorse twined on a globe superimposed on a breaking wave.

Naval Reserve Merchant Marine Insignia - A gold embroidered or metal spread eagle surcharged with crossed anchors behind a shield in the center. The letters "U.S." appear on the scroll, to the wearer's right of the shield; the letters "N.R." appear on the scroll, to the wearer's left of the shield.

Identification Badges

The Presidential Service Badge (PSB), the Vice Presidential Service Badge (VPSB), the Office of the Secretary of Defense Identification Badge (OSD ID Badge), the Joint Chiefs of Staff Identification Badge (JCS ID Badge) and the Navy Fleet/Force/ Command Badges are authorized to be worn on Navy uniforms and can be worn after detachment from qualifying duty.

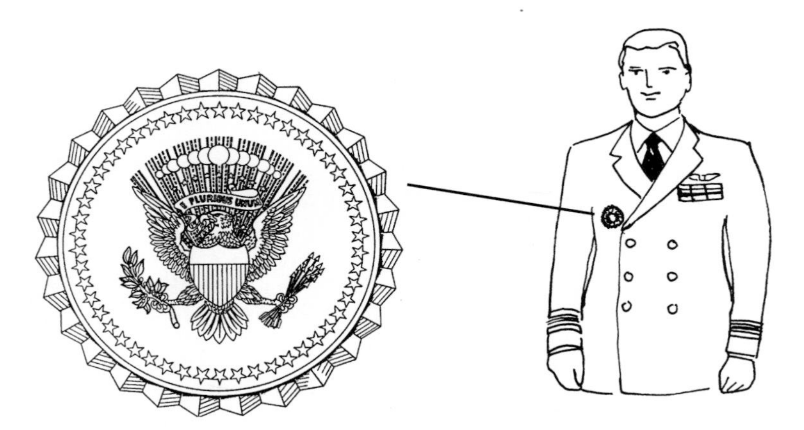

Presidential Service Badge (PSB)

The Presidential Service Badge was established on 1 September 1964 and replaced the White House Service Badge previously established on 1 June 1960. It is awarded to Naval Personnel by the Secretary of the Navy in the name of the President as permanent recognition of their contributions in the service of the President. Those eligible for the badge are members of the Armed Forces who, after 20 January 1961, have been assigned to duty in the White House or to military units and support facilities under the administration of the Military Assistant to the President for a period of at least one year. Once earned, the badge becomes a permanent part of the recipient's uniform and may be worn after the recipient leaves presidential service. The PSB consists of a blue enameled disc, 1-15/16 inches in diameter, surrounded by 27 gold rays radiating from the center. Superimposed on the disc is a gold-colored device taken from the seal of the President of the United States, encircled with 50 stars. The badge is worn on the upper right pocket during and after the period of detail.

Vice Presidential Service Badge (VPSB) (old)

Vice Presidential Service Badge (VPSB) (new)

Vice Presidential Service Badge (VPSB)

The Vice Presidential Service Badge was established on 8 July 1970. The badge is awarded in the name of the Vice President to members of the Armed Forces who have been assigned to duty in the Office of the Vice President for a period of at least one-year after 20 January 1969. The first VPSB was 1-15/16 inches overall with a white enameled disc and satin gold rays along its edge. In the center was a gold eagle with drooping wings surrounded by 50 gold stars. Once earned, the badge becomes a permanent part of the uniform. The current VPSB consists of a white enameled disc, 1-15/16 inches in diameter surrounded by 27 gold rays radiating from the center. Superimposed on the disc is a gold-colored device taken from the seal of the Vice President of the United States. The badge is worn on the upper right pocket during and after the period of detail.

Office of the Secretary of Defense Identification Badge (OSD ID)

The Office of the Secretary of Defense Identification Badge is worn by personnel who are assigned on a permanent basis to the following organizational elements:

1. Office of the Secretary and Deputy Secretary of Defense.
2. Offices of the Under Secretaries of Defense.
3. Offices of the Assistant Secretaries of Defense.
4. Office of the General Counsel of the Department of Defense.
5. Offices of the Assistants to the Secretary of Defense.
6. Office of the Defense Advisor, US Mission to the North Atlantic Treaty Organization (NATO).

After completion of one year of duty, the individual is entitled to permanent possession of the badge by issue of a certificate of eligibility. A member of the Reserve Components who is assigned an authorized Reserve Forces position in OSD for a period of no less than two years, on or after 1 January 1973, is entitled to permanent possession of the badge. The OSD ID badge consists of an eagle with wings displayed horizontally, grasping three crossed gold arrows, and having on its breast an enameled shield consisting of a blue upper portion and 13 alternating red and white stripes on the lower portion; a gold ring passing behind the wing tips bearing 13 gold stars above the eagle and a wreath of laurel and olive in green enamel below the eagle; all superimposed on a silver sunburst of 33 rays two inches in diameter.

The badge is worn on the upper left pocket during and after the period of detail.

Full Size

Miniature

Joint Chiefs of Staff Identification Badge (JCS ID)

The Joint Chiefs of Staff Identification Badge is awarded to military personnel who have been assigned to duty and served not less than one year after 13 January 1961 in a position of responsibility under the direct cognizance of the Joint Chiefs of Staff. The award of the badge must be approved by the Chairman, Joint Chiefs of Staff; the head of a Directorate of the Joint Staff; or one of the subordinate agencies of the Organization of the Joint Chiefs of Staff. Personnel are authorized to wear the badge following reassignment from duty with the JCS. The standard size JCS ID badge consists of the United States shield (upper portion in blue, and 13 stripes of alternating red and white enamel) superimposed on four gold metal unsheathed swords (two placed vertically and two diagonally), pointing to the top, with points and pommels resting on the wreath, blades and grips entwined with a gold metal continuous scroll surrounding the shield with the word JOINT at the top and the words CHIEFS OF STAFF at bottom, in blue enamel letters; all within an oval silver metal wreath of laurel 2-1/4 inches high by two inches wide. The badge is worn on the upper left pocket during and after the period of detail.

Navy Fleet/Force/Command Badges

The Navy Fleet/Force/Command Master Chief, Senior Chief and Chief Petty Officer Badges are worn by all Naval personnel serving in the capacity of Master Chief Petty Officer of the Navy, Fleet Master Chief, Force Master Chief, Command Master Chief, Command Senior Chief, or Command Chief Petty Officer per OPNAVINST 1306.2 (Series). The badges are approximately 1-3/4 inches by 1-3/8 inches, oval, and bordered with a gold chain. The Master, Senior or Chief Petty Officer device and identifying silver plate, with applicable raised silver lettering (NAVY, FLEET, FORCE, or COMMAND), are centered on a brushed gold background. Miniature badges are 1/2 size.

Navy Recruiting Command Badge

The Navy Recruiting Command Badge is worn by all Naval personnel while assigned to duty with the Navy Recruiting Command. In addition, COMNAVCRUIT-COM and COMNAVRESFOR recruiting personnel and PCN-1 (Recruit Procurement) instructors may be authorized by their respective commands to wear the Recruiting Command Identification Badge during recruiting duty or recruiter instructor duty. The badge is embossed around the outside with the words UNITED STATES NAVY and RECRUITING COMMAND with two stars. The center contains an eagle design, similar to the Bureau of Naval Personnel seal. Excellent performance meeting criteria set by COMNAVCRUIT-COM is acknowledged with the addition of a gold metallic wreath and gold or silver stars. All recruiters, officer and enlisted, serving in assigned recruiting billets in the direct recruiting production chain are eligible for the gold wreath award. Subsequent awards are signified by silver stars added to the wreath and a gold star is issued in lieu of three silver stars. A wreath is worn only while assigned to recruiting billets designated by COMNAVCRUIT-COM.

Navy Recruit Company Commander Badge

The Navy Recruit Company Commander Badge is worn by all personnel possessing an NEC of 9508 and assigned to duty as a Recruit Company Commander at a Recruit Training Command. The badge has a gold rope border with a black band inside. Inside the gold rope is embossed lettering with the words COMPANY COMMANDER. The center contains an eagle design, similar to the Bureau of Naval Personnel Seal, on a white background encircled by gold link.

Navy Career Counselor Badge

Navy Career Counselor Badge - The Navy Career Counselor Badge may be worn by all personnel in the Navy Counselor (NC) rating assigned as Command Career Counselor, all personnel possessing NEC 9588 and assigned as Command Career Counselor, and Career Information and Counseling School Instructors. The badge may also be worn by commanding officers, executive officers, and officers designated full time retention officers on the staffs of the Chief of Naval Operations, fleet commanders-in-chief, and type commanders. Those eligible to wear both Career Counselor and Navy Fleet/Force/Command Master Chief badges may only wear the latter unless specific permission to wear both has been granted by the Chief of Naval Operations. The badge has words embossed around the outside of the badge in gold lettering set in a blue background (UNITED STATES NAVY, CAREER COUNSELOR, and two stars). The center contains an eagle design, similar to the Bureau of Naval Personnel seal.

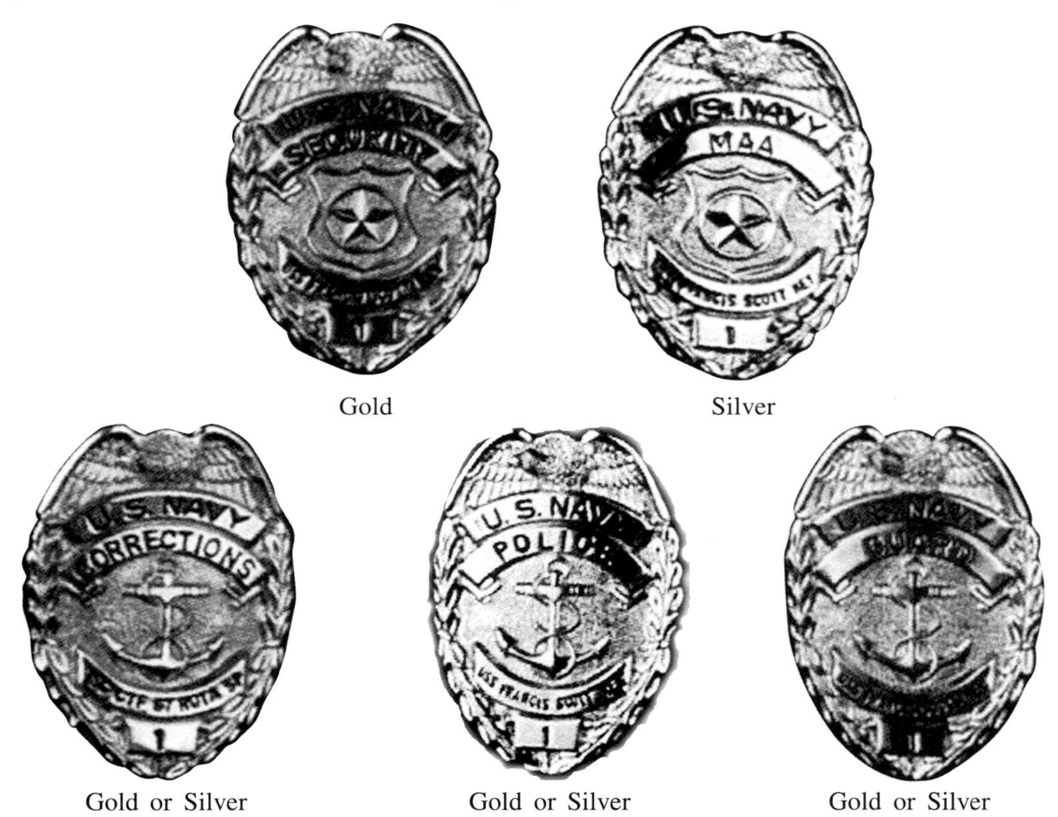

Gold · Silver

Gold or Silver · Gold or Silver · Gold or Silver

Navy Master At Arms (MAA)/Law Enforcement Badges

The MAA/Law Enforcement Badges are worn by all military personnel who are assigned to an official MAA/Law Enforcement/Physical Security or Corrections position. The badges are all the same size, but come in silver for MAA, gold for Physical Security, and gold and silver for Law Enforcement.

Joint/Unified Command Identification Badges

Joint/Unified Command Identification Badges may be authorized for Navy personnel assigned to Joint/Unified Commands may be authorized by approval from the Chief of Naval Operations (CNO). These badges are worn for the duration of assignment to that command only.

Aiguillettes, Brassards and Buttons

Aiguillettes

Aiguillettes are worn by Naval officers to identify them as aides to top-ranking government officials and flag officers. Aiguillettes are worn with both service and dress uniforms on the right shoulder by aides to the President, Vice President, foreign heads of state and White House aides. All others wear the aiguillettes on the left shoulder.

Aide to the President

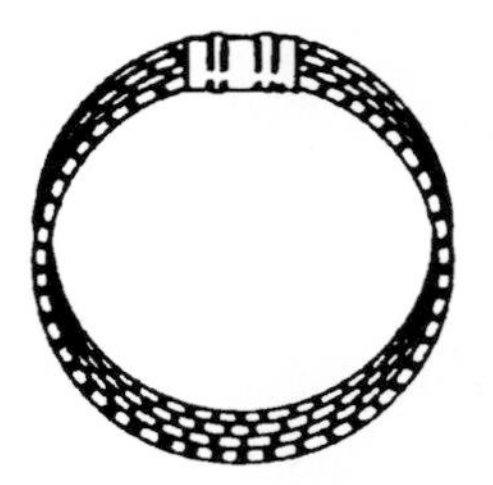

Aide to the Vice President. Aide to Admiral or official of higher rank; Naval Attaches and Assistant Naval Attaches

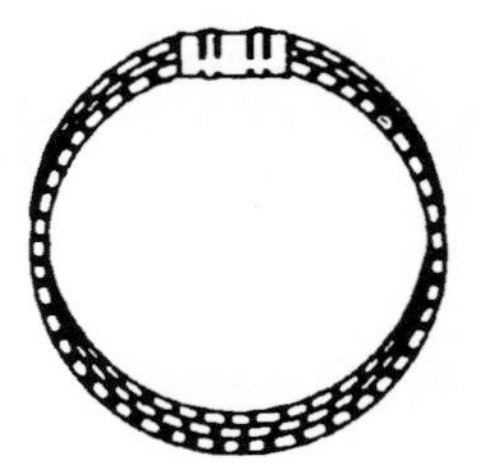

Aide to Vice Admiral

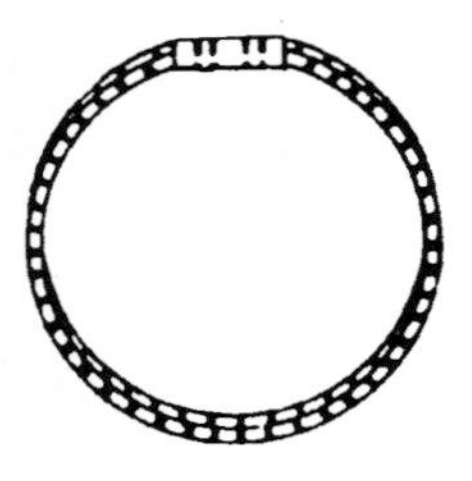

Aide to a Rear Admiral or official of lower rank; aide to a governor of a state or territory

Recruit Company Commander and Assistants

Aiguillettes are also worn by Naval personnel while serving as "A" School Military Training (ASMT) Department and Division Leading Chief Petty Officers and Company Commanders/Assistant Company Commanders; Recruit Company Commanders and their assistants; and members of the U. S. Ceremonial Guard on dress uniforms.

Service Aiguillettes

Service aiguillettes consist of a number of loops of aiguillettes cord. Service aiguillettes for Aides de Camp are cord covered with gold or gilt and other colored thread. The cord, 1/5 inch in diameter, consists of two, three, or four loops sewn together all the way around. The lengths of the cords forming loops are: The first/inside loop, 27 inches; the second loop, 28-1/2 inches; third loop, 28-3/8 inches, and fourth loop, 30-3/4 inches. Where the ends meet, the cords are fitted with a bar pin about 1-1/2 inches long by 3/8 inch wide and bound together with a 1-1/2 inch strip of No. 3 gold braid covering the ends of the cord to allow attachment of the aiguillettes to uniform coats at the shoulder, just inside the armhole seam. Presidential service aiguillettes are all gold, while other service aiguillettes for Aides are gold with dark blue spiral bands. Aiguillettes for other Naval personnel are cloth cord. Service aiguillettes consist of loops indicating:

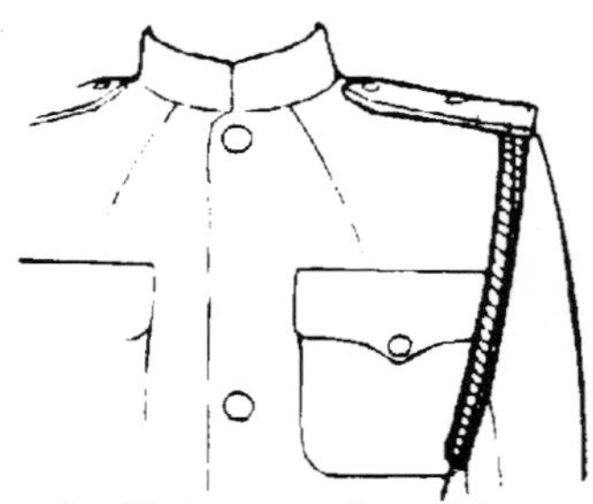

Four Loops - Personal aides to the President or Vice President; aides at the White House; aides to the Secretary or Deputy Secretary of Defense, Secretary or Under Secretary of the Navy, and Assistant Secretaries of Defense or the Navy; aide to the General Counsel of the Navy; and naval attaches and assistant attaches assigned to an embassy.

Four Loops - Aides to admirals, generals, or officials of higher grade.

Three Loops - Aides to vice admirals and lieutenant generals.

Two Loops - Aides to rear admirals, major/brigadier generals, or other officers of lower grade entitled to an aide.

Two Loops - Officers appointed as aides to a governor of a state or territory may wear aiguillettes on official occasions.

One Loop - "A" School Military Training (ASMT) Department and Division Leading Chief Petty Officers and Company Commanders/Assistant Company Commanders; Recruit Company Commanders and their assistants wear a loop of red and white. Recruit Company Commanders wear one red loop; Recruit Company Commander Assistants wear one light blue loop and members of the U. S. Ceremonial Guard wear a dark blue loop on whites and a white loop on blues.

Dress Aiguillettes

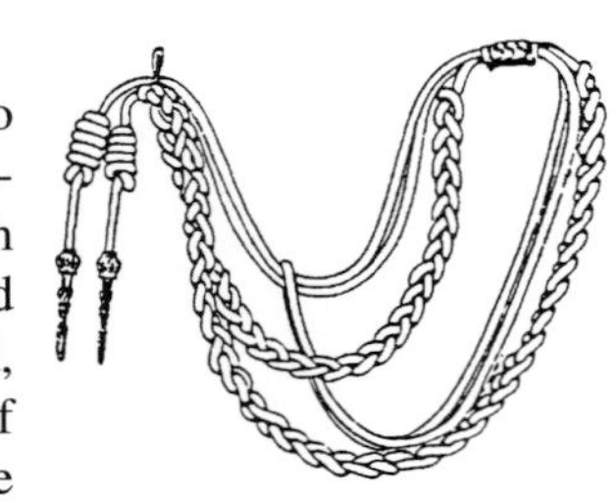

Dress aiguillettes have two single aiguillette cords, approximately 1/5 inch in diameter, with a cord of yellow cotton, covered with gold or gilt plaited thread, and two additional loops of unplaited aiguillette cord. At the termination of the plaited cords are approximately 3 inches of plain cord with two gilt metal pencils, approximately 3-1/2 inches long, fastened to the ends, and mounted with two silver anchors and a becket. Presidential service aiguillettes are all gold, while aiguillettes for other Aides have dark blue thread inserted forming plaited bands of approximately 7/16 inch.

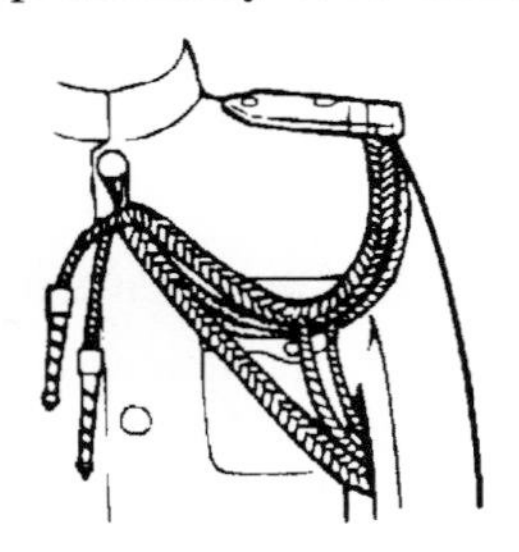

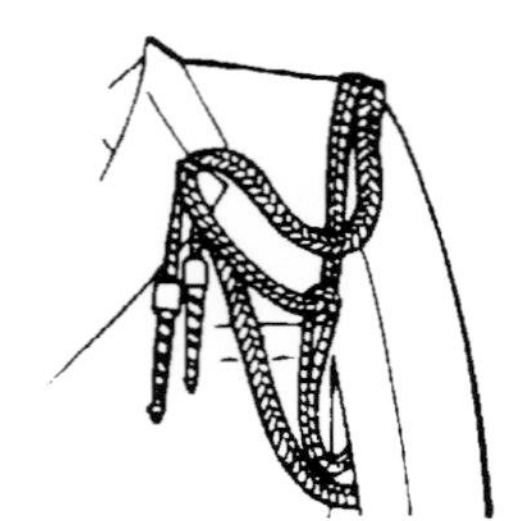

Boatswain's Pipe and Lanyard

The boatswain's pipe has been a "badge" of office for boatswains (boatswain's mates) since the early days of the Navy. The pipe and lanyard are worn around the neck while carrying out official ceremonial duties and military watches. The lanyard is braided with Belfast cord in a traditional style and sennit. When hanging free, the bottom of the pipe does not fall below the top of the belt. White lanyards are worn with dark/blue uniforms and black lanyards with white uniforms. Men place the pipe in the left breast pocket when not in use. Women place the boatswain's pipe (attached to the lanyard) between the top and second button of their service blue jacket when not in use.

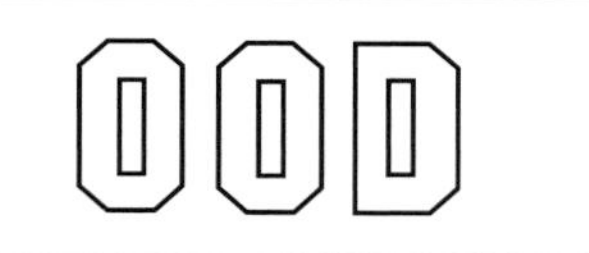

Officer of the Day (Deck)

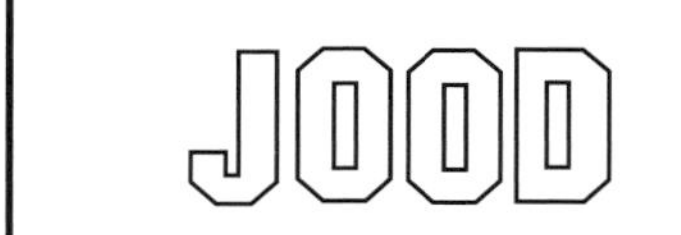

Junior Officer of the Day (Deck)

Shore Patrol

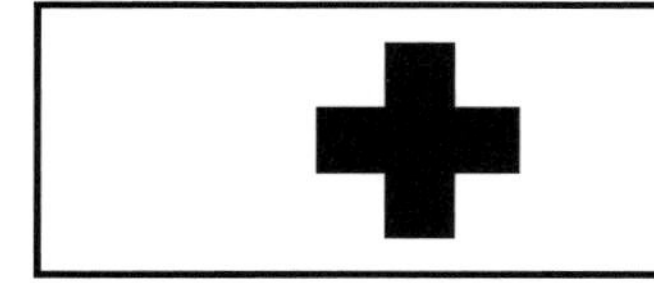

Geneva Cross

Master at Arms

Brassards

Brassards are cloth bands, marked with symbols, letters or words, indicating a type of temporary duty, to which the wearer is assigned. Brassards are worn on the right arm, midway between the shoulder and elbow, on uniforms or outer garments.

Buttons

The Navy button design consists of an eagle rising, with its wings down. The left foot is on the shank, the right foot on the stock of a plain anchor, laid horizontally, and the eagle's head faces its right. The whole is surrounded by 13 five-pointed stars and a rope. Buttons are designated in terms of "line". One line equals .025 inches, making a 40-line button one inch in diameter, and a 35-line button .875 inches in diameter.

Officers and Chief Petty Officers wear Navy eagle gold gilt (anodized) buttons. Enlisted women, E-6 and below, wear silver (anodized) buttons.

Awards and Decorations

The Navy awards program is governed by the Department of the Navy. Many of the awards and decorations are unique to the Navy, but most are common to all services.

Navy awards fall into three classifications: personal decorations and unit awards; campaign and service medals; and marksmanship badges and trophies.

Personal decorations are conferred upon individuals for acts of, heroism, acts of gallantry, meritorious service, or for personal achievement. *It should be noted that there is a term, Extraordinary Heroism "EH" (such as used in awarding the Navy Cross), which grants an enlisted member an additional 10% retirement benefit. This is a separate issue decided only by the Secretary of the Navy.*

Unit awards are awarded to entire units. The Navy participates with the other services in a system by which entire units are recognized for outstanding performance. Members of cited units are entitled to wear the appropriate award (e.g.: Navy Presidential Unit Citation, Navy Unit Commendation, etc.).

Campaign or service medals are issued to individuals who participate in particular campaigns, or periods of service for which a medal is authorized.

Marksmanship badges and trophies are awarded to individuals who demonstrate a proficiency or skill with a specific weapon during a specified practice exercise, competition or match. Marksmanship badges are worn to indicate an individual's ability and are authorized for wear on naval uniforms.

Naval personnel may wear awards described in the Navy and Marine Corps Awards Manual (SECNAVINST 1650.1).

The definition of terms according to Navy Uniform Regulations are:

Award - An all-inclusive term covering any decoration, medal, badge, or attachment bestowed on an individual.

Decoration - An award conferred on an individual for a specific act of gallantry or meritorious service.

Unit Award - An award made to an operating unit for outstanding performance and worn only by members of that unit who participated in the cited action.

Service Award - An award made to those who have participated in designated wars, campaigns, expeditions, etc., or who have fulfilled creditable, specific, service requirements.

Medal - An award issued to an individual for performance of certain duties, acts or services, consisting of a medallion hanging from a suspension ribbon of distinctive colors.

Miniature Medal - A replica of a standard size medal, made to one-half original scale. Foreign medal miniatures will not exceed the size of American miniatures. The Medal of Honor will not be worn in miniature.

Badge - An award to an individual for some special proficiency or skill, which consists of a medallion suspended from a bar or bars.

Ribbon or Ribbon Bar - A portion of the suspension ribbon of a medal, worn in lieu of the medal. Ribbon bars are also authorized for certain awards which have no medals.

Rosette - Lapel device made by gathering the suspension ribbon of the medal into a circular shape.

Lapel Pin - A miniature enameled replica of the ribbon bar.

Attachment - Any appurtenance such as a star, clasp, or other device worn on a suspension ribbon of a medal or on the ribbon bar.

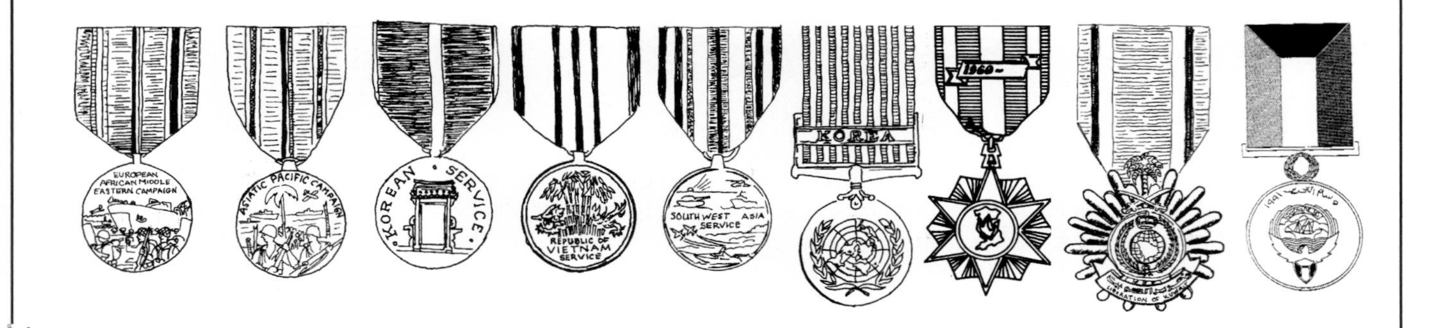

Types of Medals and Ribbons

There are two general categories of "medals" awarded by the United States to its military personnel, namely, decorations and service medals.

The terms "decoration" and "medal" are used almost interchangeably today (as they are in this book), but there are recognizable distinctions between them. Decorations are awarded for acts of gallantry and meritorious service and usually have distinctive (and often unique) shapes such as crosses or stars (there are exceptions to this, such as the Navy DSM, which is round). Medals are awarded for good conduct, participation in a particular campaign or expedition, or a noncombatant service, and normally come in a round shape. The fact that some very prestigious awards have the word "medal" in their titles (e.g.: Medal of Honor, Marine Corps Brevet Medal, Navy and Marine Corps Medal, etc.) can cause some confusion.

There are three different forms of medals (and decorations) for wear on the uniform: the full size medal, the miniature medal and the ribbon bar. The wearing of medals on the uniform is covered in the section "Wearing Medals, Ribbons and Badges" starting on page 54.

The miniature medal, the enameled lapel button or pin and the civilian hat pin may be worn on civilian clothing, and are discussed on page 55.

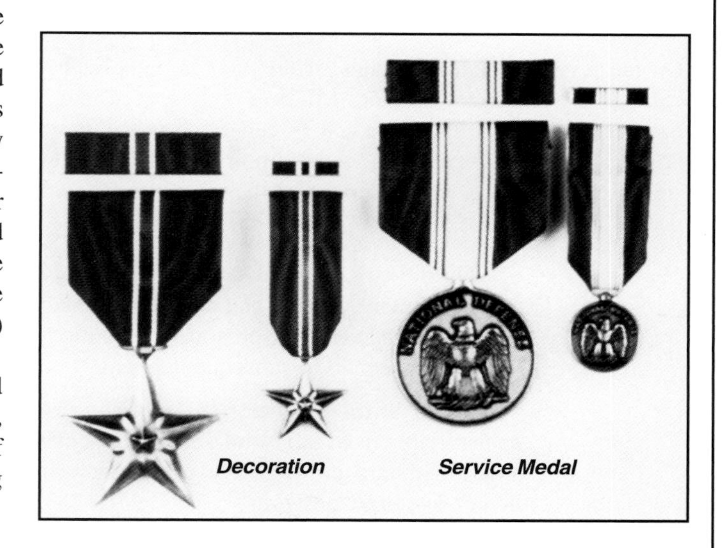

Decoration *Service Medal*

Forms of Medals

Basic Ribbon Bar

Lapel Pin

Miniature Medal

Civilian Hat pin

Full Size Medal

Reverse of Medal

Ribbon Bar with Appropriate Attachments

Attachments and Devices

Stars, clasps, numerals, letter devices, and other devices are worn on the suspension ribbons of large and miniature medals and on ribbon bars. Interestingly, each service has different regulations covering how they are worn. The wearing of attachments and devices, covered in this book, is taken from Navy Uniform Regulations and is, in many cases, unique to the Naval Service. *(The terms, attachments and devices are used interchangeably in this book.)*

In the Navy, attachment stars are worn with one ray pointing up. On large and miniature medals, if one star is authorized, it will be centered on the suspension ribbon and if more than one star is authorized, the stars will be evenly spaced in a horizontal line at the center of the suspension ribbon. On ribbon bars, if one star is authorized, it is centered, and if more than one star is authorized, the stars are evenly spaced in a horizontal line. Silver star(s) worn with bronze or gold star(s) are worn as previously mentioned, except that the first bronze star is placed to the wearer's right of the silver star(s) with additional stars alternating to the left of the silver star and so on.

Gold Stars - A gold star is worn on suspension ribbons of large and miniature medals and ribbon bars for all personal decorations in lieu of a second or subsequent award.

Silver Stars - A silver star is worn on suspension ribbons of large and miniature medals and ribbon bars in lieu of five gold stars, or in lieu of five bronze stars.

Bronze Stars - A bronze star is worn on suspension ribbons of large and miniature medals and ribbon bars to indicate a second or subsequent award or to indicate major engagements in which an individual participated.

Letter "V" - A bronze letter "V" is worn on specific combat decorations if the award is approved for valor (heroism). The specific combat decorations are; Distinguished Flying Cross, Bronze Star Medal, Air Medal, Joint Service Commendation Medal and the Navy and Marine Corps Commendation and Achievement Medals. Only one "V" is worn and gold, bronze, or silver stars, or oak leaf clusters are evenly spaced in a horizontal line with the "V" centered on the suspension ribbons of large and miniature medals. They are evenly spaced in a horizontal line on the ribbon bar with the "V" at the center.

Letter "E" - A silver letter "E" is worn on the Navy "E" ribbon. Additional awards are denoted by additional "E"s. Four or more awards are denoted by an "E" surrounded by a silver wreath.

Strike/Flight Numerals - Bronze strike/flight numerals are worn on the Air Medal to indicate the total number of Strike/Flight awards received after 9 April 1962. The Arabic numerals are placed symmetrically to the wearer's left on the suspension ribbons of large and miniature medals and to the wearer's left on the ribbon bar.

Single Mission /Individual Numerals - Gold single mission/individual numerals were worn on the Air Medal to indicate the total number of Single Mission/Individual awards received (from 1980 to 1989). The Arabic numerals were placed horizontally to the wearer's right on the suspension ribbons of large and miniature medals and ribbon bars. A bronze letter "V" (Combat Distinguishing Device) was centered to the left of the numerals on suspension ribbons of large and miniature medals and to the right on the ribbon bar.

ASIA

Navy Occupation Service Medal Clasps - The bronze Navy Occupation Service Medal clasp marked EUROPE and/or ASIA is worn on suspension ribbons of large and miniature Navy Occupation Service Medals to denote service in Europe and/or Asia.

Atlantic Fleet "A" - The block letter "A" was authorized for wear on the ribbon bar and medal suspension ribbon of the American Defense Service Medal by personnel who served in the Atlantic Fleet on the high seas prior to the outbreak of World War II.

WINTERED OVER

Antarctic Service Medal Clasp - The bronze Antarctic Service Medal clasp marked WINTERED OVER is worn on the suspension ribbon of the large Antarctic Service Medal to denote service in the Antarctic Continent during the winter months. A gold clasp is worn to denote the second winter and a silver clasp for the third. Only one clasp is worn.

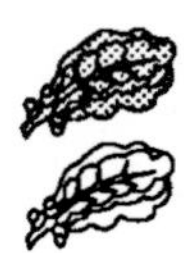

Oak Leaf Cluster - A bronze Oak Leaf Cluster denotes a second or subsequent award of a Joint Service Award bestowed upon Naval personnel by the Department of Defense. The Oak Leaf Cluster is a twig of four oak leaves and is worn on suspension ribbons of large and miniature medals and ribbon bars. The twig is worn with the stem of the oak leaves toward the wearer's right. A silver Oak Leaf Cluster is worn in lieu of five bronze Oak Leaf Clusters.

Hour Glass - A bronze hour glass device denotes ten years service on the Armed Forces Reserve Medal. Upon the completion of the ten year period, reservists that are not mobilized are awarded the Armed Forces Reserve Medal with a bronze hourglass device. Silver and gold hourglass devices are awarded at the end of twenty and thirty years of reserve service, respectively. The device represents an hourglass with the Roman numeral X superimposed thereon. It is worn centered on the suspension ribbon of the large and miniature medals and the ribbon bar.

Letter "M" - A bronze letter "M" on the Armed Forces Reserve Medal denotes reservists mobilized and called to active duty. The device is worn to the wearer's left on the suspension ribbons of large and miniature medals and the ribbon bar.

Antarctic Disc - A bronze Antarctic Disc is worn on the suspension ribbon of the medal and ribbon bar of the Antarctic Service Medal to denote wintering over on the Antarctic continent. A gold or silver disc denotes a second or third winter respectively.

Vietnam "1960-" Device - The Vietnam "1960-" Device is worn by personnel authorized to wear the Republic of Vietnam Campaign Medal. The silver banner device with the numerals "1960-" is worn on the suspension ribbon of the large and miniature medals and the ribbon bar.

Letter "W" - A silver letter "W" denotes participation in the defense of Wake Island. A bar is worn on the suspension ribbon of the medal.

Frames - A gold frame is worn on ribbon bars for some unit awards. The frame is gold plated matte finish with polished highlights. The frame is worn so that the leaves at either end will form a "V."

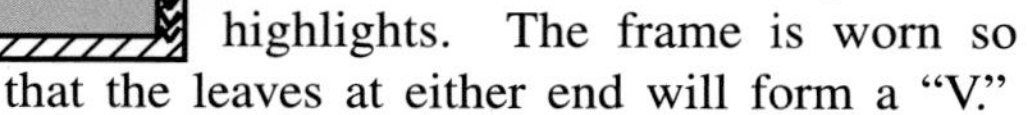

Airplane – The Berlin airlift device, a three-eighths inch gold C-54 airplane, is authorized to be worn on the ribbon bar and suspension ribbon of the Navy Occupation Service Medal by Naval and Marine Personnel who served 90 consecutive days in support of the Berlin Airlift (1948-1949).

Palm - A bronze palm is attached to the ribbon bar of the of the Republic of Vietnam Gallantry Cross Unit Citation and the Republic of Vietnam Civil Actions Unit Citation upon initial issue.

Crossed Swords and Palm Tree - The emblem of Saudi Arabia (crossed swords and a palm tree) is affixed to the center of the ribbon bar of the Kuwait Liberation Medal (Saudi Arabia).

(All attachments are shown oversized. See page 50 for placement)

Right Breast Ribbons When Wearing Medals (as seen from the front)

The Navy prescribes the wear of "ribbon-only" awards on the right breast of the full dress uniform when large medals are worn. The required display is as follows:

Joint Meritorious Unit Award	Navy Presidential Unit Citation	Combat Action Ribbon
Navy "E" Ribbon	Navy Meritorious Unit Commendation	Navy Unit Commendation
Sea Service Deployment Ribbon	Fleet Marine Force Ribbon	Reserve Special Commendation Ribbon
Navy & M/Corps Overseas Service Ribbon	Naval Reserve Sea Service Ribbon	Arctic Service Ribbon
Philippine Presidential Unit Citation	Navy Recruiting Training Service Ribbon	Navy Recruiting Service Ribbon
Vietnam Gallantry Cross Unit Citation	Vietnam Presidential Unit Citation	Korean Presidential Unit Ribbon
Pistol Marksmanship Ribbon	Rifle Marksmanship Ribbon	Philippine Independence Ribbon

Placement of Devices on Ribbons

No. of Awards	$^{3}/_{16}$ Bronze and Silver Campaign Stars	Bronze Letter V	Air Medal	
			Individual	Strike Flight
1		V		1
2		V		2
3		V		3
4		V		4
5		V		5
6				
7				
8				
9				
10				

Armed Forces Reserve Medal

After 10 years of reserve service

With 1 mobilization

With 2 mobilizations

After 10 years of reserve service and 3 mobilizations

Legend:

 = Bronze Letter "M"

 = Hourglass

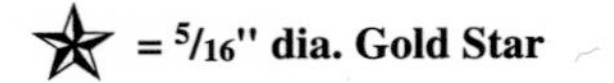

= $^{5}/_{16}$" dia. Silver Star

1, 2 etc = Bronze Block Numerals

 = Bronze Oak Leaf Cluster

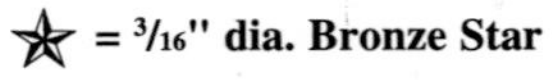

= $^{3}/_{16}$" dia. Silver Star

Placement of Devices on Medals

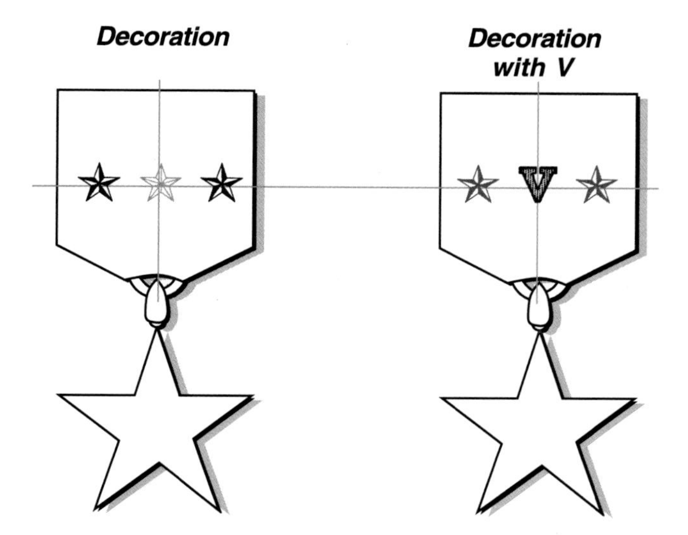

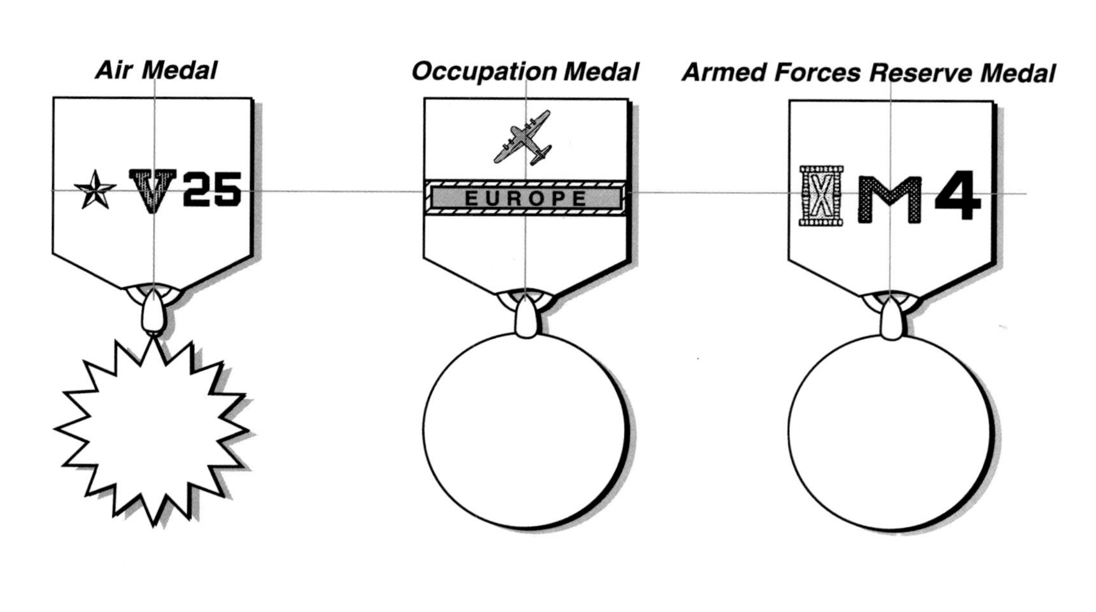

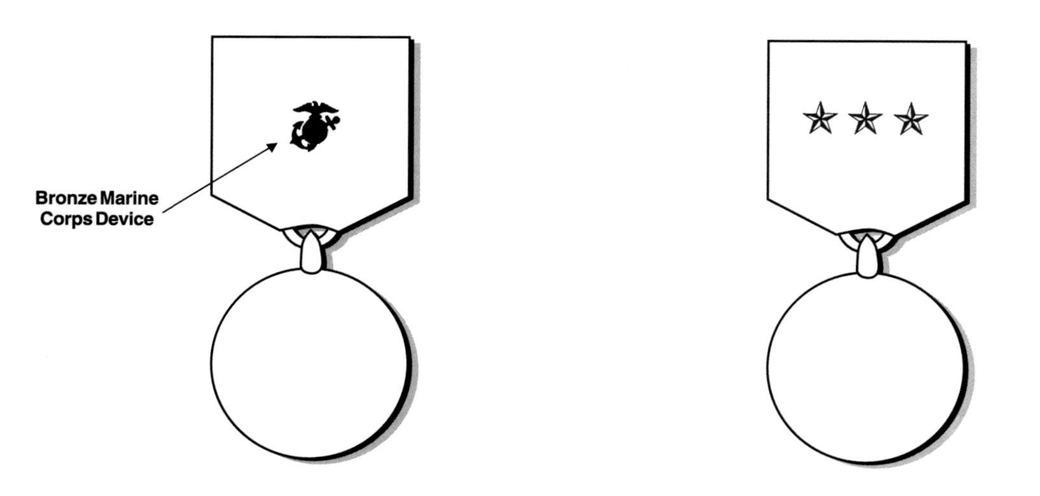

Award Certificates

U.S. Awards

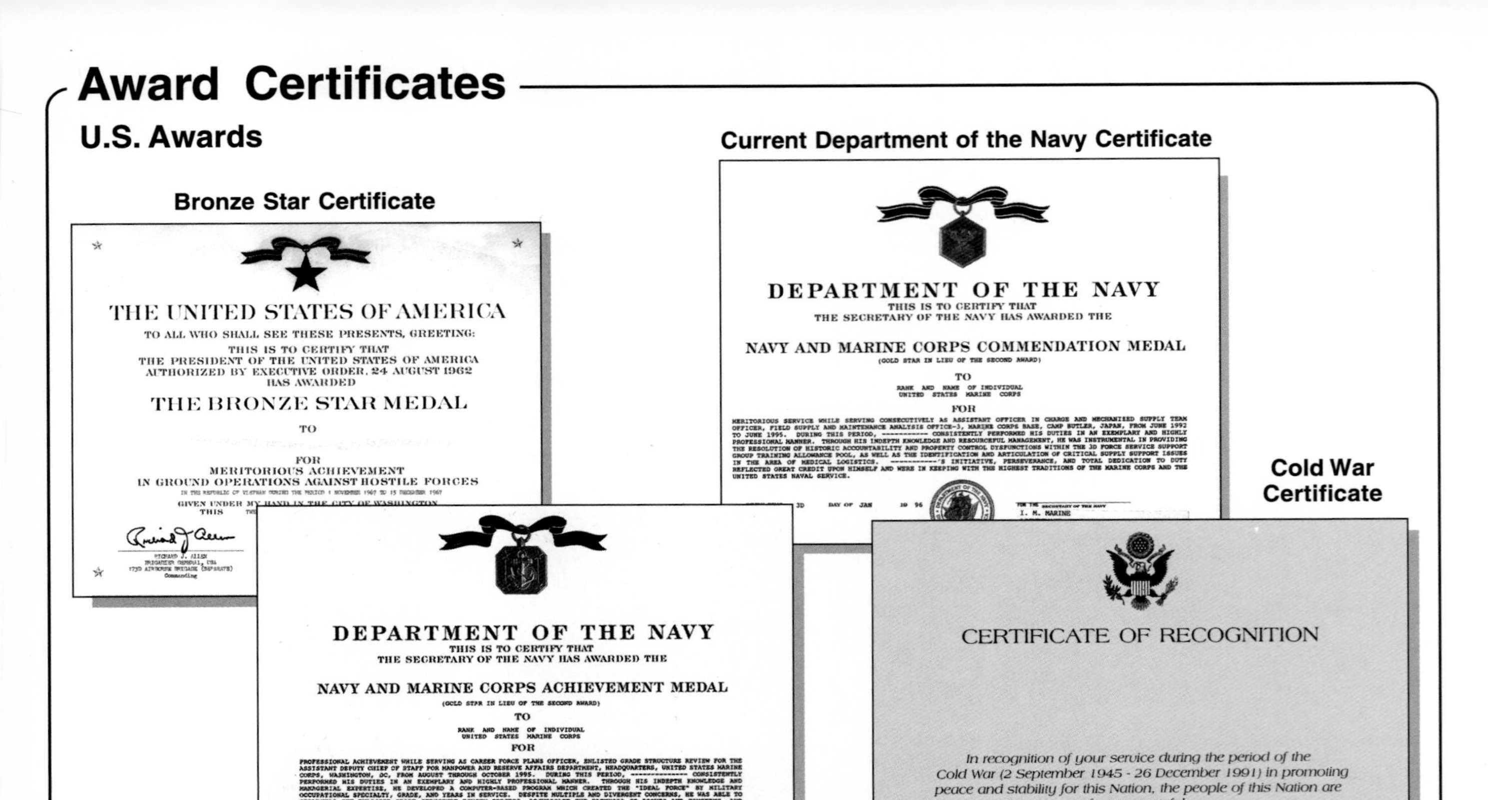

Bronze Star Certificate

Current Department of the Navy Certificate

Achievement Medal Certificate

Cold War Certificate

Foreign Awards

RVN Civil Award Certificate

Republic of Vietnam's Gallantry Cross Award Certificate and Orders

NATO Award Certificate for NATO Medal

Saudi Arabian Award Certificate for Liberation of Kuwait Medal

Claiming Medals From the U.S. Government

Veterans of any U.S. military service may request replacement of medals which have been lost, stolen, destroyed or rendered unfit through no fault of the recipient. Requests may also be filed for awards that were earned but, for any reason, were never issued to the service member. The next-of-kin of deceased veterans may also make the same request.

Requests pertaining Naval personnel should be sent to:

Navy Liaison Office (Navy Medals)
National Personnel Records Center
Room 3475/N-314
9700 Page Avenue
St. Louis, MO 63132-5100

It is recommended that requesters use Standard Form 180, Request Pertaining to Military Records, when applying. Forms are available from offices of the Department of Veterans Affairs (VA). If the Standard Form is not used, a letter may be sent, but it must include: the veteran's full name used while in the service, approximate dates of service, and service number. The letter must be signed by the veteran or his next of kin, indicating the relationship to the deceased.

It is also helpful to include copies of any military service documents that indicate eligibility for medals, such as military orders or the veteran's report of separation (DD Form 214 or its earlier equivalent). This is especially important if the request pertains to one of the millions of veterans whose records were lost in a fire at the National Personnel Records Center in 1973.

Finally, requesters should exercise extreme patience. It may take several months or, in some cases, a year to determine eligibility and dispatch medals.

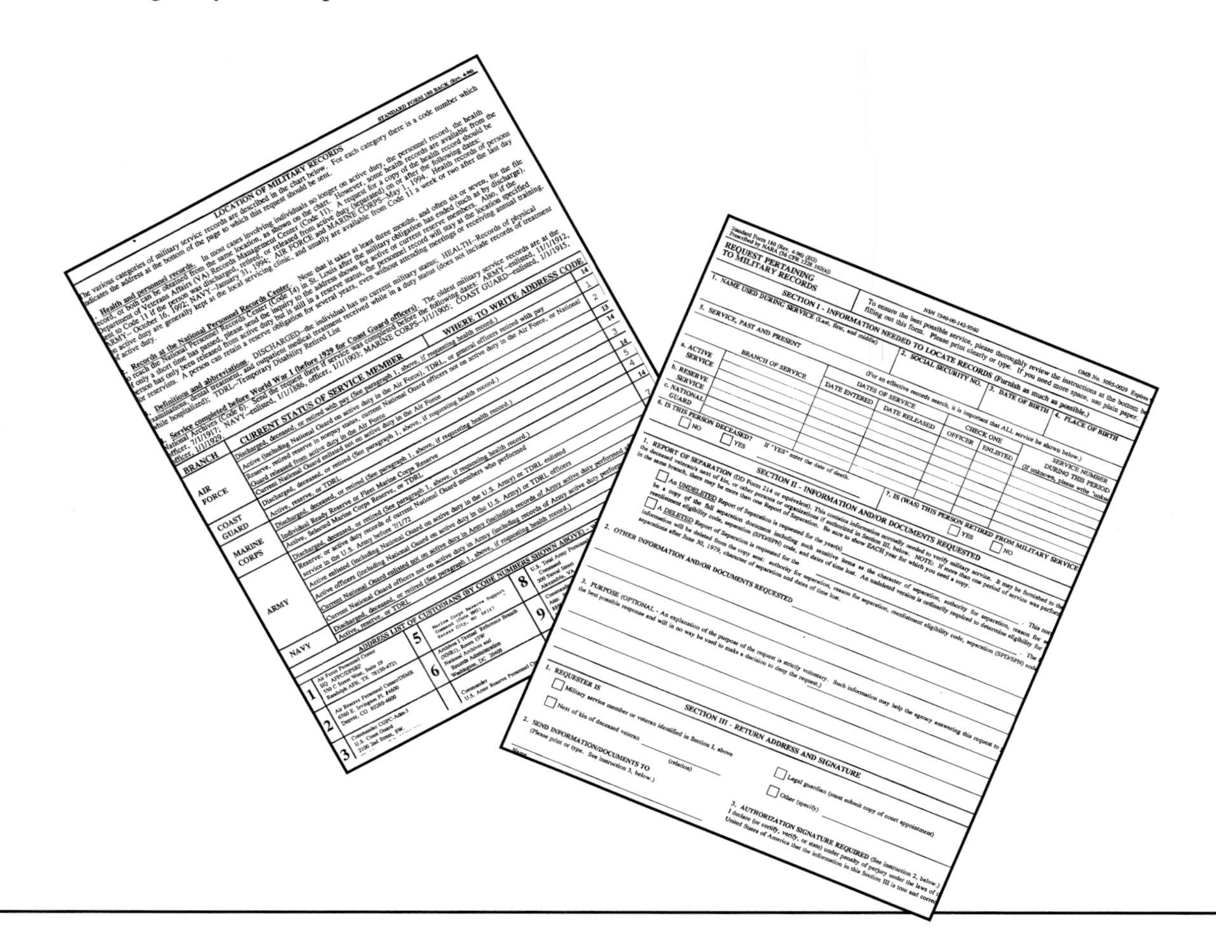

LOCATION OF MILITARY RECORDS

REQUEST PERTAINING TO MILITARY RECORDS

SECTION I - INFORMATION NEEDED TO LOCATE RECORDS

SECTION II - INFORMATION AND/OR DOCUMENTS REQUESTED

SECTION III - RETURN ADDRESS AND SIGNATURE

Wearing Ribbons and Medals

(Text taken directly from Navy Uniform Regulations)

Ribbons

Ribbons are worn on the service coat or jumper of Dress Blue and Dress White, and on the shirt of Service Khaki, Summer White, Winter Blue, and Tropical White. Ribbons are not authorized on formal dress, dinner dress, or working uniforms. Wear up to three ribbons in a single row. When more than three ribbons are authorized, wear them in horizontal rows of three each. If ribbons are not in multiples of three, the top row contains the lesser number, and the center of this row sits over the center of the one below it. Wear ribbons without spaces between ribbons or rows of ribbons. Wear ribbons with the lower edge of the bottom row centered 1/4 inch above the left breast pocket and parallel to the deck. To prevent coat lapels from covering ribbons, ribbons may be aligned so the border to wearer's left is aligned with left side of pocket. Rows of ribbons where more than 50% of the ribbon is covered by the coat lapel may contain two ribbons each and be aligned with left border.

Arrange ribbons in order of precedence in rows from top down, inboard to outboard. Wear either the three senior ribbons, or all ribbons you have earned.

Ribbons with Medals

Wear both large medals and ribbons that do not have corresponding large medals on Full Dress Uniforms. Center ribbons on the right breast in the same relative position as the holding bar of the lowest row of medals. Arrange ribbons in order of precedence in rows from top down and inboard to outboard. Wear either the senior ribbon or all ribbons. These ribbons include: Presidential Unit Citation, Navy Unit Commendation, Meritorious Unit Commendation, Navy "E" ribbon, Combat Action Ribbon, Navy Fleet Marine Force Ribbon, Sea Service Deployment Ribbon, Foreign Unit Awards, Marksmanship ribbons, etc. (see page 49). Personnel awarded only ribbons that do not have corresponding large medals shall wear the ribbons on the right side as mentioned above Do not wear ribbons on uniforms requiring miniature medals.

Large Medals

Large medals are worn on the service coat or jumper of Full Dress Blues and Full Dress White. When wearing more than one medal, suspend them from a holding bar that supports the medals' weight. Place the holding bar of the lowest row of medals in the same position as the lowest ribbon bar. The bars measure 4-1/8 inches wide, and each row of medals is 3-1/4 inches long from the top of the suspension ribbons to bottom of medals, so that bottom of medals dress in a horizontal line. Multiple rows of ribbons should be grouped with the same number of medals in each row, with the lesser number in the top row if necessary. A maximum of three medals may be worn side by side in a single row with no overlap. Arrange four or more medals (maximum of five in a single row). Overlapping shall be proportional and the inboard medal shall show in full.

Arrange medals in order of precedence in rows from top down, inboard to outboard, within rows. Service members possessing more than five medals may either wear the five senior medals or all of them.

Miniature Medals

Wear miniature medals with all formal dress uniforms and dinner dress uniforms. Each row of miniatures is 2-1/4 inches long, from top of the suspension ribbons to bottom of medals, so the bottom of medals dress in a horizontal line. Position medals so they cover the suspension ribbons of the medals in the rows below. Male officers and CPO's, and E6 and below: on formal and dinner dress jackets, place the holding bar of the lowest row of miniature medals 3 inches below the notch, centered on the lapel, parallel to the deck. On blue and white service coats, center the holding bar 1/4 inch above the left breast pocket parallel to the deck.

Female officers and CPO's, and E6 and below: on formal dress or dinner dress jackets, place the holding bar in the same relative position as on the men's dinner dress jackets, down 1/3 the distance between the shoulder seam and coat hem.

On blue and white coats, center the holding bar 1/4 inch above the left pocket flap parallel to the deck. E6 and Below: on jumper uniforms, men and women place the holding bar of the lowest row of miniature medals 1/4 inch above the pocket parallel to the deck.

Wear up to five miniature medals in a row with no overlap. (Six or more miniature medals-overlap). Arrange medals in order of precedence in rows from top down, inboard to outboard, within rows. Service members possessing five or more medals may either wear the five senior medals or all of them. On the dinner dress jacket, center up to three miniature medals on the lapel. Position four or more miniatures starting at the inner edge of the lapel extending beyond the lapel on to the body of the jacket.

Wearing of Medals, Insignia and the Uniform by Veterans and Retirees

One of the first lessons taught to new recruits is proper wear of the uniform and its insignia. The same principle applies to the wearing of military awards by veterans and retirees. There are a number of occasions when tradition, patriotism, ceremonies and social occasions call for the wear of military awards.

Civilian Dress

Purple Heart Ribbon Lapel Pin

The most common manner of wearing a decoration or medal is as a lapel pin in the left lapel of a civilian suit jacket. The small enameled lapel pin represents the ribbon bar of a single decoration or ribbon an individual has received (usually the highest award or one having special meaning to the wearer). Many well-known veterans such as former Senator Bob Dole, a World War II Purple Heart recipient, wear a lapel pin. Pins are available for all awards and some ribbons such as the Combat Action Ribbon or the Navy Presidential Unit Citation.

Naval Aviator Lapel Pin

A small Navy insignia and miniature wings are also worn in the lapel or as a tie tac. Additionally, retirees are encouraged to wear their Navy retired lapel pin and World War II veterans are encouraged to wear their Honorable Discharge lapel pin (affectionately referred to as the "ruptured duck").

WW2 Honorable Discharge Lapel Pin

Retirement Lapel Pin

Formal Civilian Wear

For more formal occasions, it is correct and encouraged to wear miniature decorations and medals. For a black or white tie occasion, the rule is quite simple: if the lapel is wide enough, wear the miniatures on the left lapel or, in the case of a shawl lapel on a tuxedo, the miniature medals are worn over the left breast pocket. The center of the holding bar of the bottom row of medals should be parallel to the deck immediately above the pocket. Do not wear a pocket handkerchief.

Uniform

On certain occasions retired Navy personnel may wear either the uniform prescribed at the date of retirement or any of the current active duty authorized uniforms. Retirees should adhere to the same grooming standards as Navy active duty personnel when wearing the uniform (for example, a beard is currently inappropriate while in uniform). Whenever the uniform is worn, it must be done in such a manner as to reflect credit upon the individual and the Navy. (Do not mix uniform items.)

The occasions for wear by retirees are:

- military ceremonies.
- military funerals, weddings, memorial services and inaugurals.
- patriotic parades on national holidays.
- military parades in which active or reserve units are participating.
- educational institutions when engaged in giving military instruction or responsible for military discipline.
- social or other functions when the invitation has obviously been influenced by the members having at one time been on active service.

Honorably separated wartime veterans may wear the uniform authorized at the time of their service.

The occasions are:

- military funerals, memorial services, and inaugurals.
- patriotic parades on national holidays.
- any occasion authorized by law.
- military parades in which active or reserve units are participating.

Non-wartime service personnel separated (other than retired and Reserve) are not authorized to wear the uniform but may wear their medals.

Color Plates

The following color plates provide examples of the decorations, medals, ribbons, badges and insignia used in the United States Navy since World War II. All of these are manufactured in the United States to a very exacting standard.

Page 57 displays examples of uniform insignia, which includes insignia for both cap and collar. The larger cap insignia are worn on combination caps, while the smaller versions are worn on garrison caps. The Naval Reserve Merchant Marine insignia is worn on the Merchant Marine uniform by Merchant Marine officers who are also in the U.S. Naval Reserve.

Pages 58 through Page 60 displays examples of metal breast insignia. Breast insignia denote command responsibility; a warfare qualification or a special qualification designation. Most breast insignia are worn on the left breast pocket above ribbons or medals. The command insignia (at-sea and ashore) shown on page 59 are worn on the right breast of the service uniform when incumbent and on the left breast, post tour. The large breast insignia are worn on service uniforms, while the smaller versions are worn on dress uniforms, when miniature medals are prescribed. Officers' insignia are normally gold colored, while enlisted insignia are normally silver colored. It should be noted that some of these insignia are now obsolete.

Page 61 displays identification badges and a civilian blazer badge. The identification badges are worn on all uniforms other than working uniforms. The Presidential and Vice Presidential service badges are worn on the right, while the other identification badges are worn on the left. A miniature version of these identification badges are worn on shirt pockets.

Page 62 displays metal rank and rate insignia for commissioned, warrant and petty officers. Also shown are the lapel anchors and collar insignia for the Naval Academy and N.R.O.T.C. Midshipmen.

Page 63 displays staff corps devices for commissioned and warrant officers. The Buddhist Chaplain and Bandleader insignia are the embroidered variety worn on the sleeve of the dress blue uniform and the shoulder boards of the tropical and summer white uniforms. All the other insignia are the metal variety worn on the left collar points of the working uniforms. It should be noted that the insignia for the Civil Engineer Corps is the same for both commissioned and warrant officers.

Page 64 displays examples of a variety of cloth insignia.

Top row: Several shoulder boards, including Rear Admiral (Lower Half), Captain (Line), Commander (Nurse Corps), Lieutenant Commander (Medical Service Corps), Lieutenant (Supply Corps), Lieutenant Junior Grade (Dental Corps), Ensign (Supply Corps) and a Chief Warrant Officer W-2 (Physician's Assistant and Technical Nurse).

Second row: Three soft shoulder boards worn with the white shirt (Nurse Corps, Dental Corps, Technical Nurse and Master Chief). In the center is a Midshipman Commander's shoulder board. To the right are Chief Petty Officer's dress blue sleeve insignia (Electrician's Mate and Electronics Technician).

Third row: Rating badges for Petty Officer First, Second and Third Class Petty (Pipe Fitter, Aviation Electrician's Mate and Construction Mechanic) as well as service stripes (the gold stripes indicate 16 years of good conduct service).

Fourth row: A Petty Officer First Class (female) rating badge (Navy Counselor), Petty Officer First Class dungaree rating badge and Dentalman group rate with striker mark. In the center are utility uniform collar devices for personnel assigned to Marine Corps (Fleet Marine Force) units (top - Hospital Corpsman; bottom - Dental Technician). At the far right are the good conduct service stripes worn on the dress white uniform.

Bottom row: E-3 group rate with striker mark (Gunners Mate, Machinist Mate, Aviation Electronics Technician, Utilitiesman, Sonar Technician, Engineman, Aviation Electrician's Mate, and Steelworker.). At the far right are rating badges (top - Airman, center - Fireman Apprentice and bottom - Seaman Recruit).

Pages 65 through 70 display, in detail, the medals and ribbons of the United States Navy. These awards are displayed in the correct order of precedence. Examples of foreign decorations and commemorative medals are also displayed. Descriptions of these awards and award criteria are detailed in the section immediately following the color plates.

Page 71 displays examples of appropriate ways to display Navy insignia and awards. There are examples from World War II, Korea, Vietnam, Kuwait and peacetime service.

Uniform Insignia

Officer Insignia
(Garrison Cap)

Officer Insignia
(Combination Cap)

Naval Reserve Merchant
Marine Insignia

MCPO of the Navy
(Combination Cap)

Master Chief Petty Officer
(Combination Cap)

Senior Chief Petty Officer
(Combination Cap)

Chief Petty Officer
(Combination Cap)

Collar Insignia

Warrant Officer (W-1)
(Combination Cap)

Midshipman
(Combination & Garrison Cap)

Women (E1-E6)
(Combination Cap)

Petty Officer
1st Class

Petty Officer
2nd Class

Petty Officer
3rd Class

Breast Insignia

Naval Astronaut Aviator (Miniature)

Naval Astronaut Flight Officer

Naval Aviator

Balloon Pilot

Naval Flight Officer

Flight Surgeon

Naval Aviation
(Supply Corps)

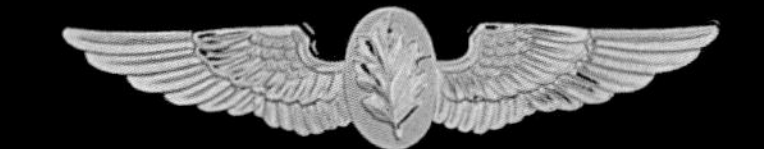

Flight
Nurse

Aviation
Psychologist

Naval Aviation Observer

Aircrew

Combat Aircrew

Aviation Warfare Specialist

Submarine (Officer)

Submarine (Enlisted)

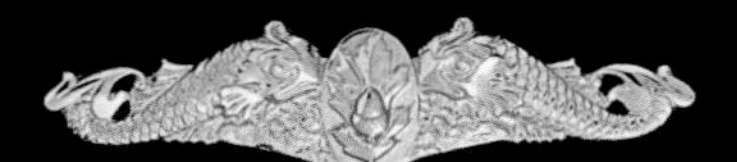

Submarine
(Medical)

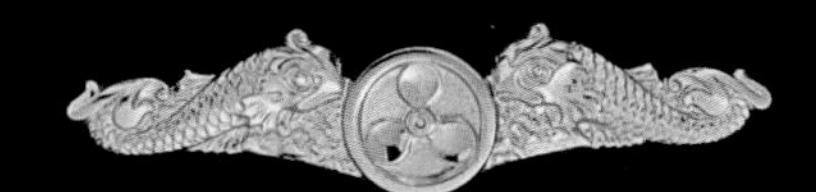

Submarine
(Engineering Duty)

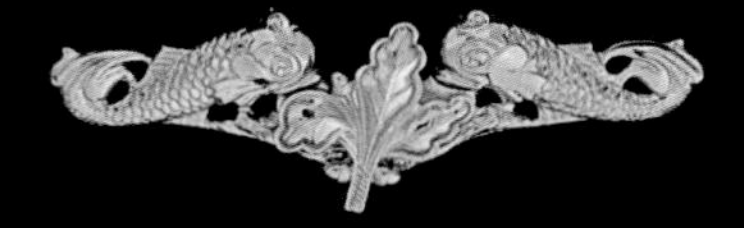

Submarine
(Supply Corps)

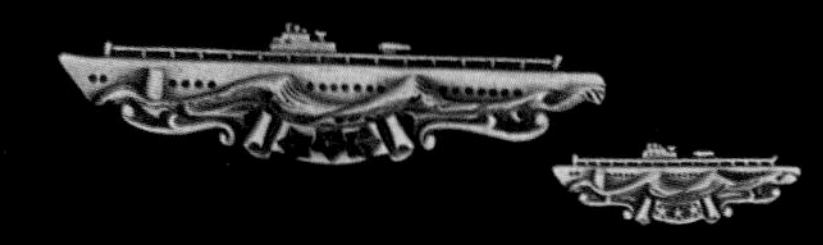

Submarine Combat Patrol (Miniature)

Deterrent Patrol

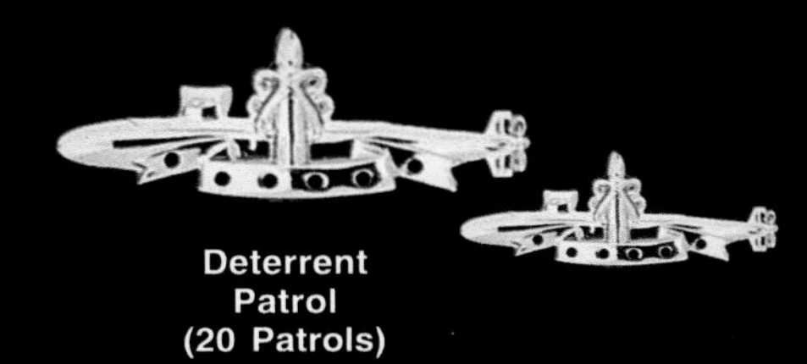

Deterrent Patrol (20 Patrols)

Surface Warfare (Officer)

Surface Warfare (Enlisted)

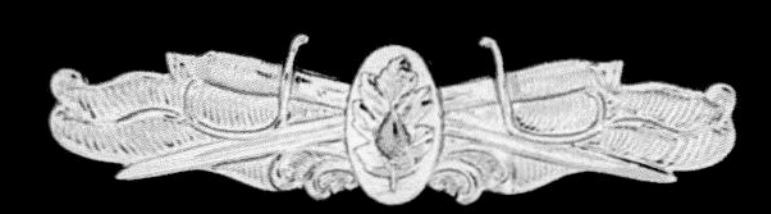

Surface Warfare (Medical Corps)

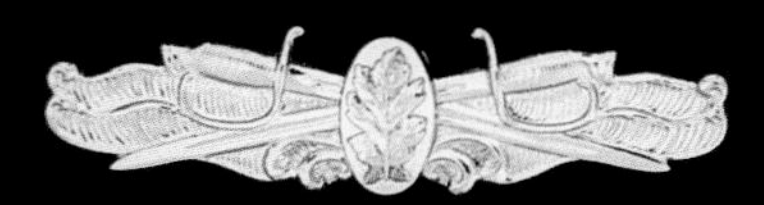

Surface Warfare (Dental Corps)

Surface Warfare (Medical Surface Corps)

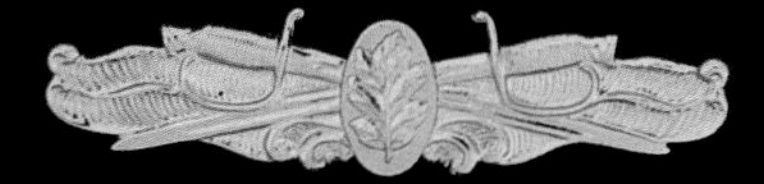

Surface Warfare (Nurse Corps)

Surface Warfare (Supply Corps)

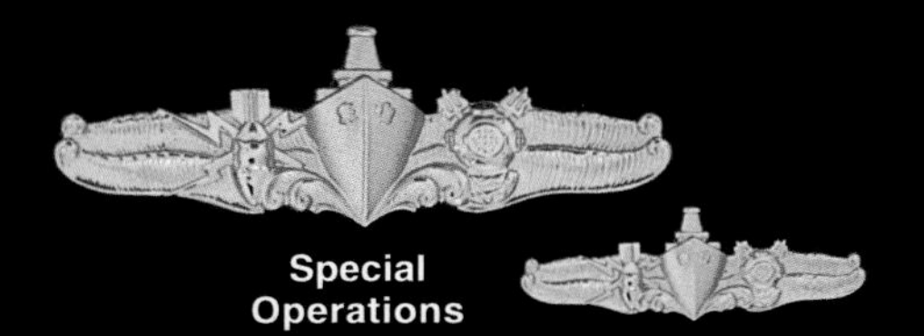

Special Operations

Command At Sea

Command Ashore/Project Mgr

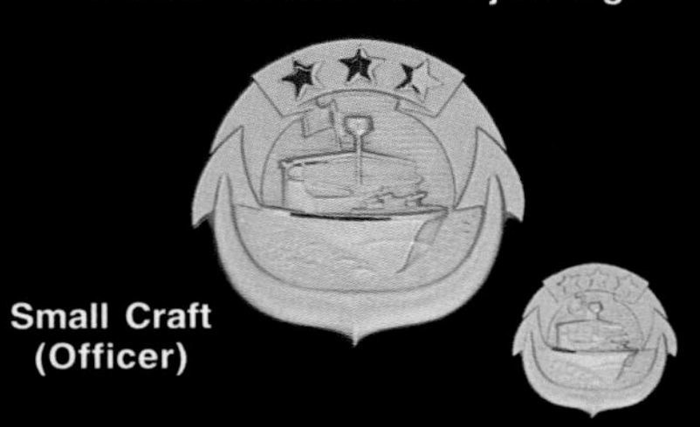

Small Craft (Officer)

Small Craft (Enlisted)

Craft Master

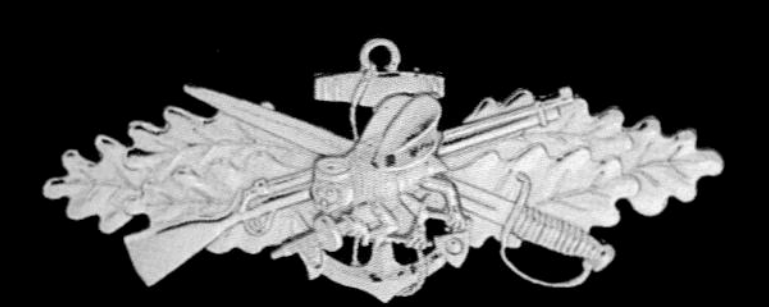

Seabee Combat Warfare (Officer)

Seabee Combat Warfare (Enlisted)

Breast Insignia

Special Warfare (SEAL)

Special Warfare (SEAL)
Enlisted (Obsolete)

Underwater
Demolition Team
(UDT Officer)
(Obsolete)

Underwater
Demolition Team (UDT
Enlisted) (Obsolete)

Explosive
Ordnance Disposal
(EOD)

Senior Explosive
Ordnance Disposal
(EOD)

Master Explosive
Ordnance Disposal
(EOD)

Integrated Undersea
Surveillance

Basic Parachutist

Naval
Parachutist

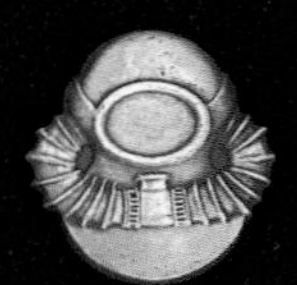
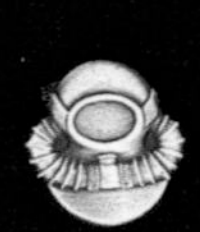

Scuba Diver

Second Class
Diver

First Class Diver

Diving Officer

Master Diver

Diving (Medical)

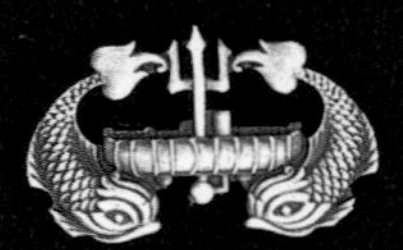

Deep Submergence

Presidential Service

Vice Presidential Service

Secretary of Defense

Joint Chiefs of Staff

Navy Recuiting Command

Navy Career Counselor

(Miniature)

Civilian Blazer

Master at Arms

Navy Master Chief Badge

Fleet Master Chief Badge

Force Master Chief Badge

Command Master Chief Badge

Command Senior Chief Badge

Command Chief Badge

Rank and Rate Insignia

Officer

Fleet Admiral

Admiral

Vice Admiral

Rear Admiral

Rear Admiral (Lower Half) Commodore

Captain

Commander

Lt. Commander

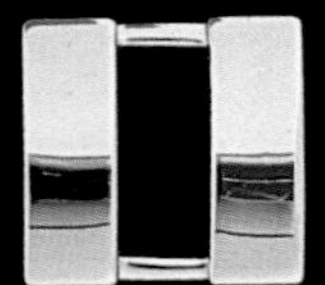

Lieutenant

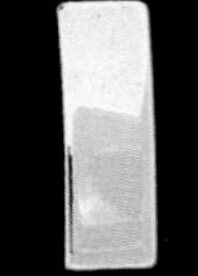

Lieutenant (JG)

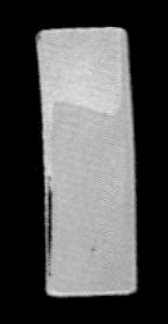

Ensign

Warrant Officer

CWO-4

CWO-3

CWO-2

WO-1

MCPO-Navy

MCPO

SCPO

CPO

PO 1C

PO 2C

PO3C

Midshipman

Lapel Insignia

Naval Academy and N.R.O.T.C. Collar Insignia

Mid. Capt.

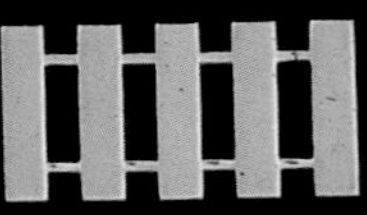

Mid. Cdr.

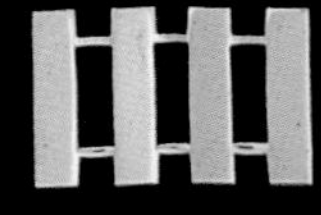

Mid. Lt. Cdr.

Mid. Lt.

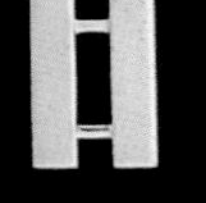

Mid. Lt.(JG)

Mid. Ens.

Mid. PO

Line and Staff Corps Devices

Officer

Medical Corps

Nurse Corps

Medical Service Corps

Dental Corps

Supply Corps

JAG Corps

Law Community

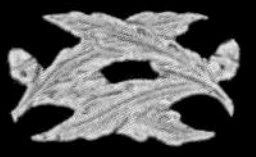
Civil Engineer Corps (same for Warrant Officer)

Buddhist Chaplain

Jewish Chaplain

Muslim Chaplain

Christian Chaplain

Bandleader

Warrant Officer

Aerographer

Air Traffic Control Technician

Aviation Boatswain

Aviation Electronics Technician

Aviation Maintenance Technician

Aviation Operations Technician

Aviation Ordnance Technician

Boatswain

Communication Technician

Cryptologic Technician

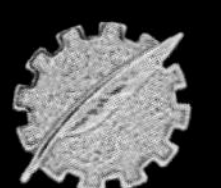
Data Processing Technician

Diving Officer

Electronics Technician

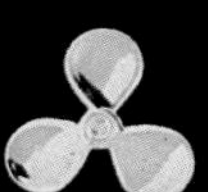
Engineering Technician

Explosive Ordnance Technician

Intelligence Technician

Operations Technician

Ordnance Technician

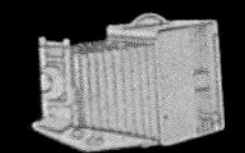
Photographer

Security Technician

Repair Technician

Ship's Clerk

Underwater Ordnance Technician

Physician's Assistant

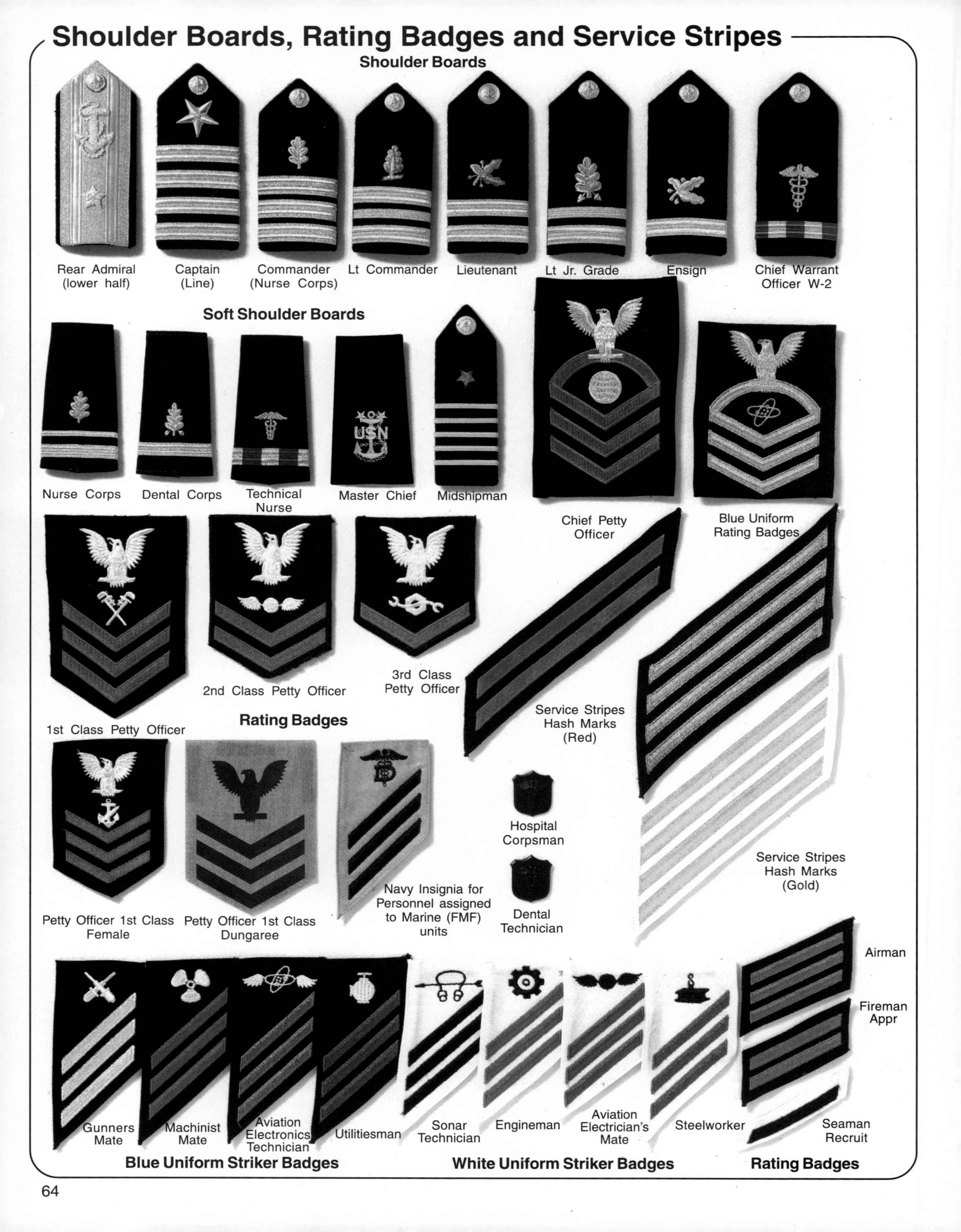
Shoulder Boards, Rating Badges and Service Stripes
Shoulder Boards
Rear Admiral (lower half)
Captain (Line)
Commander (Nurse Corps)
Lt Commander
Lieutenant
Lt Jr. Grade
Ensign
Chief Warrant Officer W-2
Soft Shoulder Boards
Nurse Corps
Dental Corps
Technical Nurse
Master Chief
Midshipman
Chief Petty Officer
Blue Uniform Rating Badges
1st Class Petty Officer
2nd Class Petty Officer
3rd Class Petty Officer
Rating Badges
Service Stripes Hash Marks (Red)
Petty Officer 1st Class Female
Petty Officer 1st Class Dungaree
Navy Insignia for Personnel assigned to Marine (FMF) units
Hospital Corpsman
Dental Technician
Service Stripes Hash Marks (Gold)
Airman
Fireman Appr
Seaman Recruit
Gunners Mate
Machinist Mate
Aviation Electronics Technician
Utilitiesman
Sonar Technician
Engineman
Aviation Electrician's Mate
Steelworker
Blue Uniform Striker Badges
White Uniform Striker Badges
Rating Badges

**Navy
Medal of Honor
(1861-1913)**
Pg. 74

**Navy
Medal of Honor
(1913-1917)**

**Navy
Medal of Honor
(1917-1942)**
Pg. 74

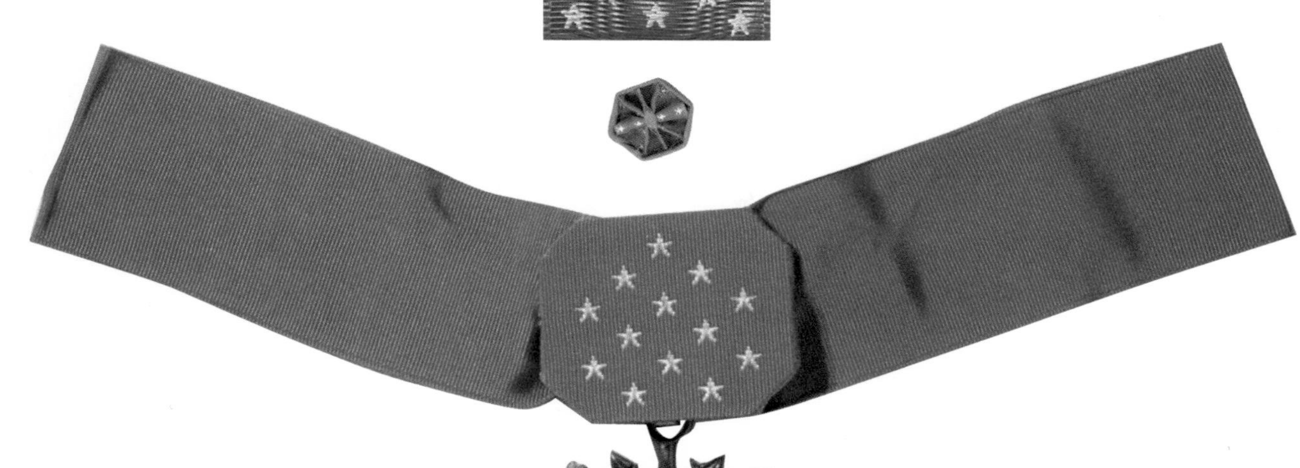

**Navy
Medal of Honor
(Current)**
Pg. 74

U.S. Personal Decorations

Navy Cross
Pg. 75

Defense Distinguished
Service Medal Pg. 75

Navy Distinguished
Service Medal Pg. 76

Silver Star
Pg. 76

Defense Superior
Service Medal Pg. 77

Legion of Merit
Pg. 77

Distinguished
Flying Cross Pg. 78

Navy & Marine Corps
Medal Pg. 79

Bronze Star
Pg. 79

Purple Heart
Pg. 80

Defense Meritorious
Service Medal Pg. 81

Meritoriuos Service Medal
Pg. 81

U.S. Personal Decorations and Service Medals

Air Medal
Pg. 82

Joint Service
Commendation Medal Pg. 83

Navy & Marine Corps
Commendation Medal Pg. 83

Joint Service
Achievement Medal Pg. 84

Navy & Marine Corps
Acheivment Medal Pg. 84

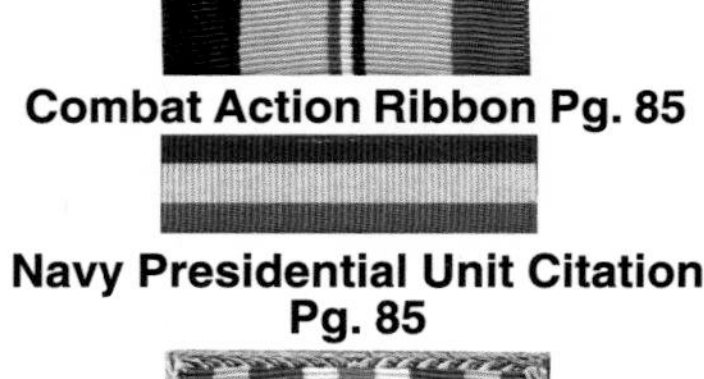

Combat Action Ribbon Pg. 85

Navy Presidential Unit Citation
Pg. 85

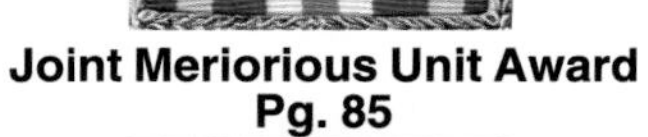

Joint Meriorious Unit Award
Pg. 85

Navy Unit Commendation Pg.86

Navy Meritorious Unit Commendation Ribbon P. 86

Navy "E" Ribbon Pg. 86

Prisoner Of War
Medal Pg. 87

Navy Good Conduct
Medal Pg. 87

Reserve Special
Commendation Ribbon
Pg. 88

Fleet Marine Force
Ribbon
Pg. 89

U.S. Antarctic
Expedition Medal Pg. 97

Naval Reserve Meritorious
Service Medal Pg. 89

Navy Expeditionary Medal
Pg. 90

China Service Medal
Pg. 90

U.S. Service Medals

American Defense
Service Medal Pg. 91

American Campaign
Medal Pg. 91

European-African-Middle
Eastern Campaign Medal
Pg. 92

Asiatic-Pacific
Campaign Medal Pg. 93

World War II
Victory Medal Pg. 94

Navy Occupation Service
Medal Pg. 94

Medal for
Humane Action Pg. 95

National Defense
Service Medal Pg. 95

Korean Service Medal
Pg. 96

Antarctica Service Medal
Pg. 97

Armed Forces
Expeditionary Medal Pg. 98

Vietnam Service Medal
Pg. 99

Southwest Asia Service Medal Pg. 100

Kosovo Campaign Medal Pg. 100

Armed Forces Service Medal Pg. 101

Humanitarian Service Medal Pg. 101

Outstanding Volunteer Service Medal Pg. 102

Navy Sea Service Deployment Ribbon Pg. 102

Navy Arctic Service Ribbon Pg. 102

Naval Reserve Sea Service Ribbon Pg. 103

Navy & Marine Corps Overseas Service Ribbon Pg. 103

Navy Recruiting Service Ribbon Pg. 103

Navy Recruit Training Service Ribbon Pg. 104

Armed Forces Reserve Medal Pg. 104

Naval Reserve Medal (obsolete) Pg. 104

Philippine Presidential Unit Ctation Pg. 105

Vietnam Presidential Unit Citation Pg. 106

Foreign Decorations Pg. 105

Korean Presidential Unit Citation Pg. 105

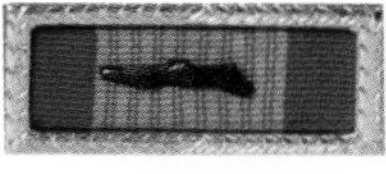
Vietnam Gallantry Cross Unit Citation Pg. 106

Vietnam Civil Actions Unit Citations Pg. 106

Philippine Defense Medal Pg. 107

Philippine Liberation Medal Pg. 107

Philippine Independence Medal Pg. 108

Foreign Decorations and Marksmanship Awards

United Nations Service Medal (Korea) Pg. 108

United Nations Medal (Observer Medal) Pg. 109

NATO Medal for Bosnia Pg. 109

NATO Medal for Kosovo Kosov Pg. 112

Multinational Force and Observers Medal Pg. 112

Inter-American Defense Board Medal Pg. 113

Republic of Vietnam Campaign Medal Pg. 113

Saudi Arabian Medal for the Liberation of Kuwait Pg. 113

Kuwaiti Medal for the Liberation of Kuwait Pg. 114

Republic of Korea War Service Medal Pg. 114

Expert Rifleman Medal Pg. 115

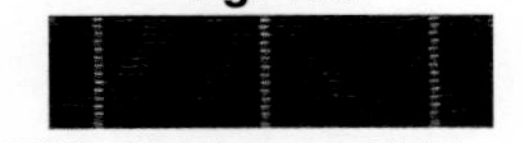

Rifle Marksman Ribbon Pg.115

Expert Pistol Shot Medal Pg. 115

Pistol Marksman Ribbon Pg. 115

Examples of Award Displays

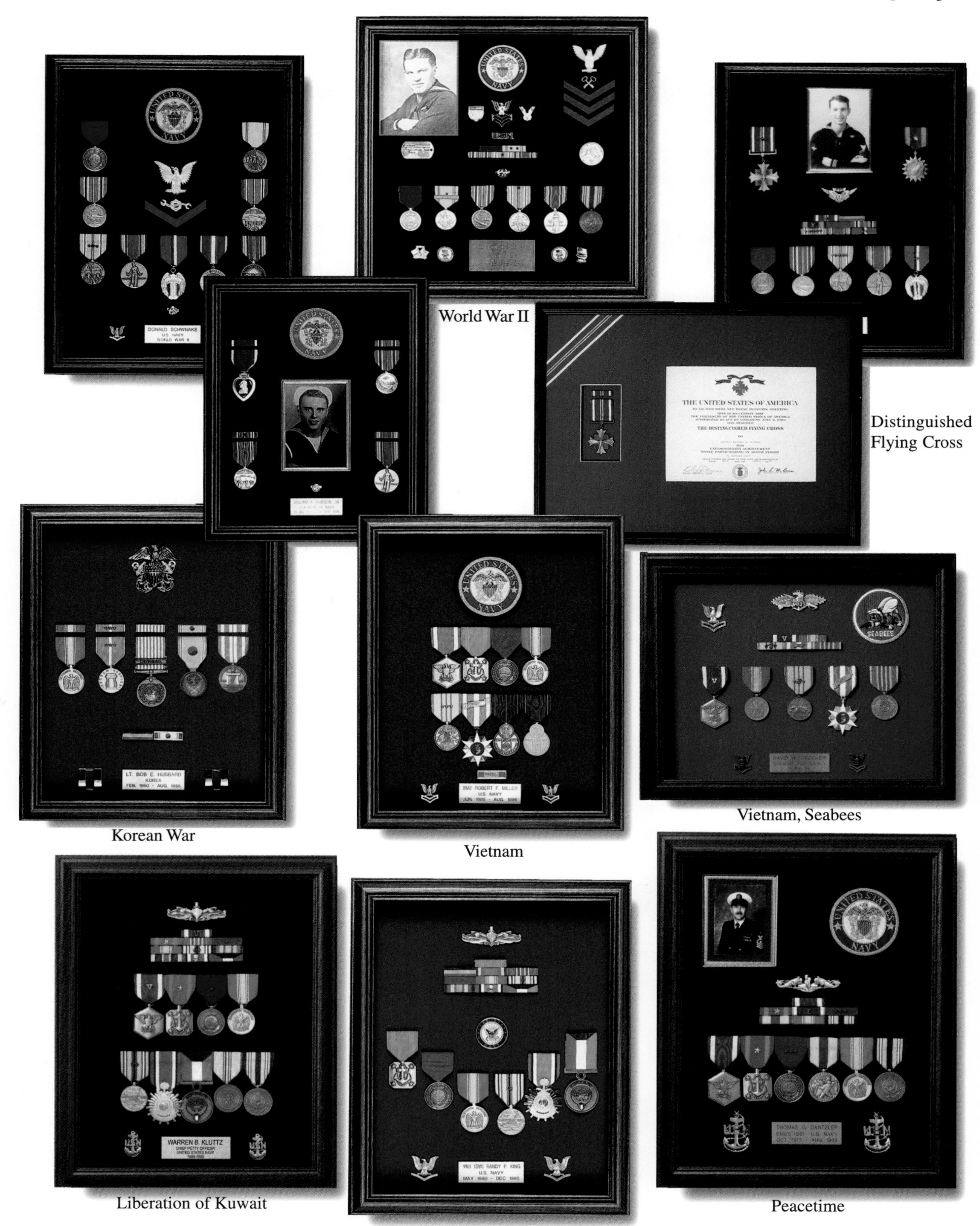

World War II

Distinguished Flying Cross

Korean War

Vietnam

Vietnam, Seabees

Liberation of Kuwait

Peacetime

Ribbon Precedence Chart

U.S. Navy

Correct Order Of Ribbon Wear (Left Breast)

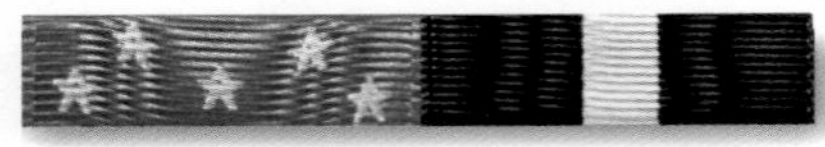

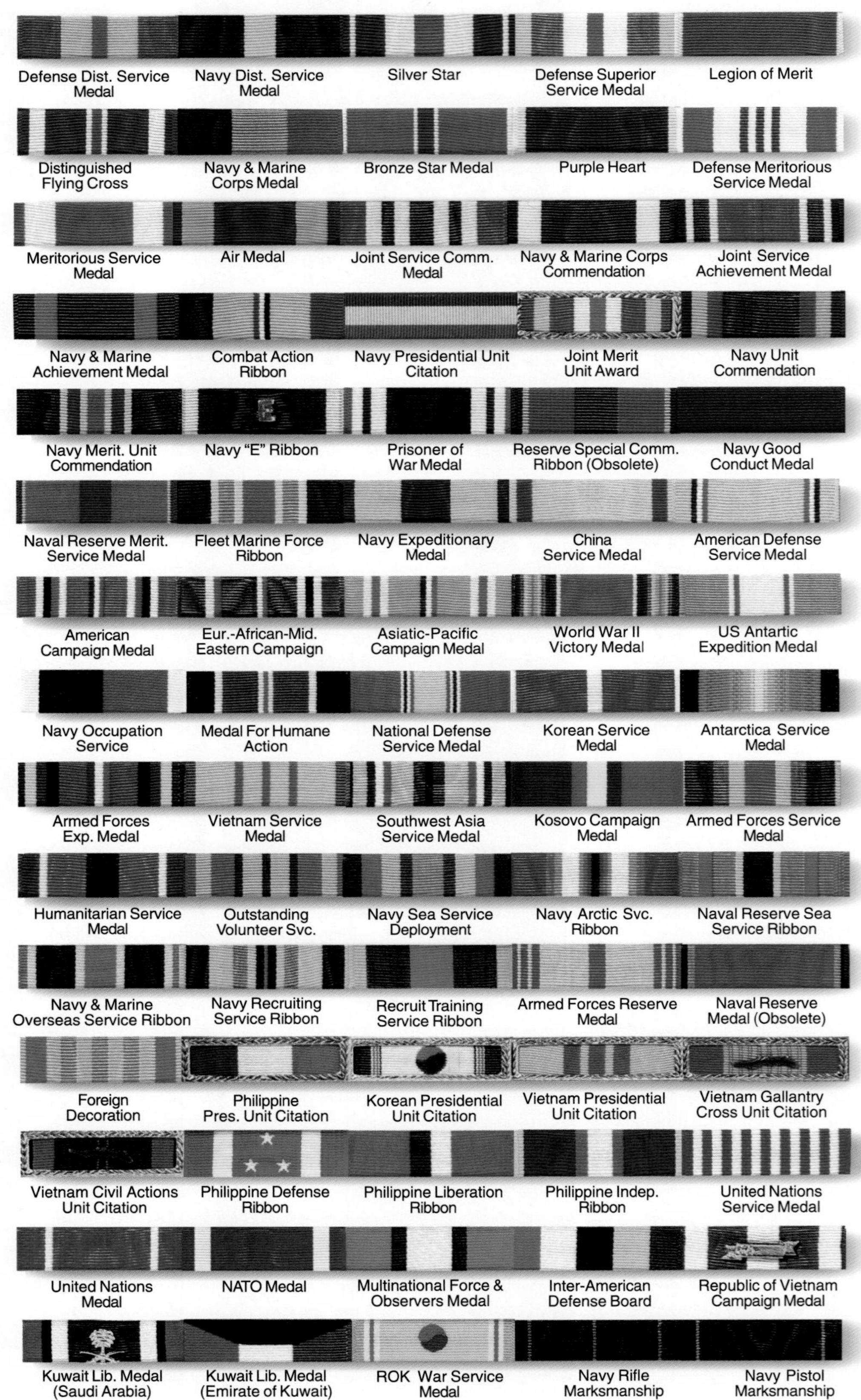

Note: Per Navy regulations, no row may contain more than three (3) ribbons. The above display is arranged solely to conserve space on the page.

Ribbon Devices

Medal of Honor	Navy Cross
Gold	Silver Gold

Defense Distinguished Service Medal Bronze	Navy Distinguished Service Medal Silver Gold	Silver Star Silver Gold	Defense Superior Service Medal Silver Bronze	Legion of Merit Bronze Silver Gold
Distinguished Flying Cross Bronze Silver Gold	Navy & Marine Corps Medal Gold	Bronze Star Medal Bronze Silver Gold	Purple Heart Silver Gold	Defense Meritorious Service Medal Silver Bronze
Meritorious Service Medal Silver Gold	Air Medal Bronze Silver Gold	Joint Service Commendation Medal Bronze Silver Bronze	Navy & Marine Corps Commendation Medal Bronze Silver Gold	Joint Service Achievement Medal Silver Bronze
Navy & Marine Corps Achievement Medal Bronze Silver Gold	Combat Action Ribbon Silver Gold	Navy Presidential Unit Citation Gold Gold Silver Bronze	Joint Meritorious Unit Award Silver Bronze	Navy Unit Commendation Silver Bronze
Navy Meritorious Unit Commendation Silver Bronze	Navy "E" Ribbon Silver Silver	Prisoner of War Medal Silver Bronze	Reserve Special Commendation Ribbon (Obsolete) None	Navy Good Conduct Medal Silver Bronze
Naval Reserve Meritorious Service Medal Silver Bronze	Fleet Marine Force Ribbon None	Navy Expeditionary Medal Silver Silver Bronze	China Service Medal Bronze	American Defense Service Medal Bronze Bronze
American Campaign Medal Bronze	European-African-Middle Eastern Campaign Bronze Silver Bronze	Asiatic-Pacific Campaign Medal Bronze Silver Bronze	World War II Victory Medal None	US Antartic Expedition Medal
Navy Occupation Service Gold Airplane	Medal For Humane Action None	National Defense Service Medal Bronze	Korean Service Medal Bronze Silver Bronze	Antarctica Service Medal Bronze, Gold, or Silver
Armed Forces Expeditionary Medal Bronze Silver Bronze	Vietnam Service Medal Bronze Silver Bronze	Southwest Asia Service Medal Bronze Bronze	Kosovo Campaign Medal Bronze Bronze	Armed Forces Service Medal Silver Bronze
Humanitarian Service Medal Silver Bronze	Outstanding Volunteer Service Medal Silver Bronze	Navy Sea Service Deployment Ribbon Silver Bronze	Navy Arctic Service Ribbon None	Naval Reserve Sea Service Ribbon Silver Bronze
Navy & Marine Corps Overseas Service Ribbon Silver Bronze	Navy Recruiting Service Ribbon Silver Bronze	Recruit Training Service Ribbon Silver Bronze	Armed Forces Reserve Medal Bronze, Silver, Gold Hourglass Bronze	Naval Reserve Medal (Obsolete) None
Foreign Decoration As Specified by the Awarding Government	Philippine Presidential Unit Citation Bronze	Republic of Korea Presidential Unit Citation None	Republic of Vietnam Presidential Unit Citation None	Vietnam Gallantry Cross Unit Citation Bronze Palm
Vietnam Civil Actions Unit Citation Bronze Palm	Philippine Defense Ribbon Bronze	Philippine Liberation Ribbon Bronze	Philippine Independence Ribbon None	United Nations Service Medal None
United Nations Medal Bronze	NATO Medal Bronze	Multinational Force & Observers Medal Bronze Numeral	Inter-American Defense Board Medal Gold	Republic of Vietnam Campaign Medal Silver Date Bar
Kuwait Liberation Medal (Saudi Arabia) Gold Palm Tree	Kuwait Liberation Medal (Emirate of Kuwait) None	ROK War Service Medal	Navy Rifle Marksmanship Ribbon Silver Bronze	Navy Pistol Marksmanship Ribbon Silver Bronze

Note: Per Navy regulations, no row may contain more than three (3) ribbons.
The above display is arranged solely to conserve space on the page.

= Gold Star
Denotes second and subsequent awards of a decoration

= Silver Star
Worn in the same manner as the gold star, in lieu of five gold stars

= Bronze Service Star
Denotes second and subsequent awards of a service award or participation in a campaign or major operation

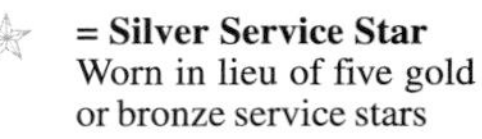
= Silver Service Star
Worn in lieu of five gold or bronze service stars

= Bronze Oak Leaf Cluster
Denotes second and subsequent awards of a Joint Service decoration or unit citation

= Silver Oak Leaf Cluster
Worn in lieu of five bronze oak leaf clusters

= Bronze Letter "V"
Awarded for distinguished actions in combat (valor)

= Letter "A"
Denotes service with Atlantic Fleet prior to World War II

= Bronze Letter "M"
Denotes reservists mobilized and called to active duty

= Antarctica Disk
Denotes personnel who "winter-over" on the Antarctic continent

= Marine Corps Device
Denotes combat service with Marine Corps Units

= Bronze Numeral
Denotes total number of strike/flight awards of the Air Medals and other awards

Marksmanship Devices

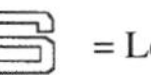
= Letter "S"

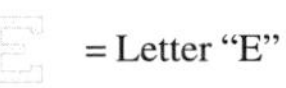
= Letter "E"

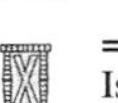
= Hourglass
Issued after each 10 years of reserve service

MEDAL OF HONOR
(Navy-Marine Corps-Coast Guard Design)

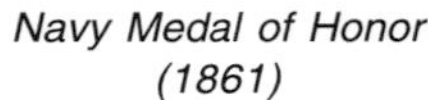

Navy Medal of Honor (1861)

Navy Medal of Honor "Tiffany Cross" (1919-1942)

Navy Medal of Honor (1942-Present)

"For conspicuous gallantry and intrepidity at the risk of life, above and beyond the call of duty, in action, involving actual conflict with an opposing armed force." The Medal of Honor is worn before all other decorations and medals and is the highest honor that can be conferred on a member of the Armed Forces. Since its inception, 3,432 Medals of Honor have been awarded to 3,408 individuals.

The Medal of Honor was signed into law by President Lincoln on 21 December 1861. This Public Resolution 82 contained a provision for a Navy medal of valor. At first the decoration was intended to recognize gallantry in action by enlisted personnel, but was later amended to include officers. Congress also passed an act on 9 July 1918 which established criteria for the award that the act of heroism had to be above and beyond the call of duty and so unique as to clearly distinguish the recipient from his comrades.

In 1919, the so called "Tiffany Cross" Medal of Honor version came into use and remained until 1942, when the current Navy Medal of Honor was instituted. The1919 version was often referred to as the "Tiffany Cross," since Tiffany was involved with its design. The medal was a gold cross patee on a wreath of oak and laurel leaves. In the center of the cross was an octagon with the inscription UNITED STATES NAVY 1917-1919. Inside the octagon was an eagle design of the United States Seal and an anchor appeared on each arm of the cross. The reverse of the medal had the raised inscription AWARDED TO and space for the recipient's name. The medal was suspended by a light blue ribbon with thirteen white stars. The ribbon was suspended from a rectangular gold pin bar inscribed with the words VALOUR.

Many Americans today are confused by the term "Congressional Medal of Honor," when, in fact, the proper term is "Medal of Honor." A law passed in July 1918 authorized the President to present the medal in the name of Congress. Part of this confusion stems from the fact that all MOH recipients belong to the Congressional Medal of Honor Society chartered by Congress.

An act of Congress in July 1963 clarified and amended the criteria for awarding the Medal of Honor to prevent award of the medal for deeds done "in line of profession," but not necessarily in actual conflict with an enemy. This act of Congress made the clarification by stating that the award was "for service in military operations involving conflict with an opposing force or for such service with friendly forces engaged in armed conflict."

A recommendation for the Navy Medal of Honor must be made within three years from the date of the deed upon which it depends, and award of the medal must be made within five years after the date of the deed. A stipulation for the medal is that there must be a minimum of two witnesses to the deed, who swear separately that the event transpired as stated in the final citation.

The current Navy Medal of Honor has been revised three times and has been in its current form since August 1942. The medal is a five pointed star with a standing figure of the Goddess Minerva surrounded by a circle of stars representing the number of States in the Union at the outbreak of the Civil War. Minerva, the Goddess of Strength and Wisdom, holds a shield taken from the Great Seal of the United States, and in her left hand she holds a fasces, which represents the lawful authority of the state; she is warding off a crouching figure representing discord. The medal is suspended from an anchor and the reverse is plain for engraving the recipient's name. The ribbon is light blue and has an eight-sided central pad with thirteen white stars.

NAVY CROSS

Silver Gold

Instituted: 1919
Criteria: Extraordinary heroism in action against an enemy of the U.S. while engaged in military operations involving conflict with an opposing foreign force or while serving with friendly foreign forces
Devices: Gold, silver star
Notes: Originally issued with a 1-1/2" wide ribbon

For extraordinary heroism (not justifying the Medal of Honor) in connection with military operations against an opposing armed force. The Navy Cross is worn after the Medal of Honor and before all other decorations.

The Navy Cross was established by an Act of Congress and approved on 4 February 1919. Initially the Navy Cross was awarded for extraordinary heroism or distinguished service in either combat or peacetime. The criteria was upgraded in August 1942 to limit the award to those individuals demonstrating extraordinary heroism in connection with military operations against an armed enemy.

The Navy Cross medal is a cross patee with the ends of the cross rounded. It has four laurel leaves with berries in each re-entrant angle, which symbolizes victory. In the center of the cross is a sailing ship on waves. The ship is a caravel, symbolic of ships of the fourteenth century. On the reverse are crossed anchors with cables attached with the letters USN amid the anchors. The ribbon is navy blue with a white center stripe. Additional awards of the Navy Cross are denoted by gold stars five-sixteenths of an inch in diameter.

DEFENSE DISTINGUISHED SERVICE MEDAL

Bronze Silver

Instituted: 1970
Criteria: Exceptionally meritorious service to the United States while assigned to a Joint Activity in a position of unique and great responsibility
Devices: Bronze, silver oak leaf cluster

Awarded by the Secretary of Defense for exceptionally meritorious service in a duty of great responsibility. The Defense Distinguished Service Medal is worn after the Navy Cross and before the Navy Distinguished Service Medal.

The Defense Distinguished Service Medal was established by an Executive Order, which was signed by President Nixon on 9 July 1970. The medal was instituted for senior officers who held positions of authority over elements of other service branches. This eliminated the need to award multiple Distinguished Service Medals from the service branches involved. The medal is the highest award for meritorious service within the Department of Defense.

The Defense Distinguished Service Medal was designed by the Army's Institute of Heraldry and is gold in color featuring a blue enameled pentagon superimposed by a gold eagle with outspread wings. On the eagle's breast is the seal of the United States and in its talons are three arrows. The eagle and pentagon are surrounded by a gold circle consisting of thirteen stars and laurel and olive branches. At the top of the circle are five gold rays extending above the stars, which form the medal suspension. On the reverse of the pentagon is the raised inscription FROM THE SECRETARY OF DEFENSE TO, with a space for inscribing the recipient's name. On the reverse of the gold circle is the raised inscription FOR DISTINGUISHED SERVICE. The ribbon has a narrow red center stripe of flanked on either side by stripes of white, blue and gold. Additional awards of the Defense Distinguished Service Medal are denoted by oak leaf clusters.

NAVY DISTINGUISHED SERVICE MEDAL

Silver

Gold

Instituted: 1919
Criteria: Exceptionally meritorious service to the U.S. Government in a duty of great responsibility
Devices: Gold, silver star
Notes: 107 copies of earlier medal design issued but later withdrawn. First ribbon design was 1-1/2" wide

For exceptionally meritorious service to the Government in a duty of great responsibility. The Navy Distinguished Service Medal is worn after the Defense Distinguished Service Medal and before the Silver Star.

The Navy Distinguished Service Medal was established by an Act of Congress and approved on 4 February 1919 and, like the Navy Cross, was made retroactive to 6 April 1917. During this period there was confusion about what criteria constituted the award of the Navy Distinguished Service Medal and what criteria constituted the award of the Navy Cross. At the outbreak of World War II, laws governing the award of naval decorations were changed with Public Law 702, which placed the Navy Cross above the Navy Distinguished Service Medal and clearly limited the Navy Distinguished Service Medal for exceptionally meritorious service and not for acts of heroism. The very first Navy Distinguished Service Medal was awarded, posthumously, to Brigadier General Charles M. Doyen, USMC.

The Navy Distinguished Service Medal is a gold medallion with an American bald eagle with displayed wings in the center. The eagle is surrounded by a blue enameled ring which contains the words, UNITED STATES OF AMERICA with NAVY at the bottom. Outside the blue ring is a gold border of waves. The medal is suspended from its ribbon by a five pointed white enameled star with an anchor in the center. Behind the star are gold rays emanating from the re-entrant angles of the star. The reverse of the medal contains a trident surrounded by a wreath of laurel. The wreath is surrounded by a blue enamel ring with the inscription FOR DISTINGUISHED SERVICE. The blue enamel ring is surrounded by a gold border of waves the same as on the front of the medal. The ribbon is navy blue with a gold stripe in the center. Additional awards of the Navy Distinguished Service Medal are denoted by gold stars five sixteenths of an inch in diameter. The Navy Distinguished Service Medal is the highest peacetime award.

SILVER STAR

Silver

Gold

Instituted: 1932
Criteria: Gallantry in action against an armed enemy of the United States or while serving with friendly foreign forces
Devices: Bronze, silver oak leaf cluster
Notes: Derived from the 3/16" silver "Citation Star" previously worn on Army campaign medals

For gallantry in action (1) against an enemy of the United States; (2) while engaged in military operations involving conflict with an opposing foreign force; or, (3) while serving with friendly foreign forces engaged in armed conflict with an opposing foreign force in which the United States is not a belligerent party. The required gallantry, while of a lesser degree than required for the award of the Navy Cross, must nevertheless have been performed with marked distinction. The Silver Star is worn after the Navy Distinguished Service Medal and before the Defense Superior Service Medal.

The Silver Star was originally established by President Woodrow Wilson on 9 July 1918 as a three-sixteenth inch silver citation star to be worn on the World War I Victory Medal to denote receipt of a special letter of commendation. Although the commendation star was widely used during World War I, it was not a popular award. Arguments centered around the fact that the award was "insignificant in size and constitutes very little tangible evidence of gallantry, is not an article which can be handed down to posterity and, therefore, serve as evidence of a grateful nation and people with attendant stimulation to patriotism." Because of these arguments, the Army decided to redesign the citation star by placing it on a medal. The medal as it is today was created in 1932 for the Army and extended to the Navy and Marine Corps on 8 August 1942.

The medal was designed by Bailey, Banks and Biddle, an Atlanta retail jeweler. The medal is a five pointed gilt-bronze star with a small silver star centered in the middle, which is actually a representation of the original citation star. The small silver star is surrounded by a laurel wreath with rays radiating outward from the star. The reverse of the medal has the raised inscription FOR GALLANTRY IN ACTION and room for inscribing the recipient's name. The ribbon, employing the colors of the American Flag, has a wide red center stripe flanked on either side by a wide white stripe, a wide dark blue stripe, a thin white stripe and a thin dark blue stripe at the edges. Additional awards are denoted by five-sixteenth inch diameter gold stars.

DEFENSE SUPERIOR SERVICE MEDAL

Instituted: 1976
Criteria: Superior meritorious service to the United States while assigned to a Joint Activity in a position of significant responsibility
Devices: Bronze, silver oak leaf cluster

Awarded by the Secretary of Defense for superior meritorious service in a position of significant responsibility while assigned to a joint activity. The Defense Superior Service Medal is worn after the Silver Star and before the Legion of Merit.

The Defense Superior Service Medal was established by an Executive Order, which was signed by President Ford on 6 February 1976. The medal was instituted to recognize duty performed with distinction and significance that would justify an award comparable to the Legion of Merit for members of the Armed Forces assigned to the Office of the Secretary of Defense and other activities in the Department of Defense.

The Defense Superior Service Medal was designed by the Army's Institute of Heraldry and is of the same design as the Defense Distinguished Service Medal. The medal is silver in color featuring a blue enameled pentagon superimposed by a gold eagle with outspread wings. On the eagle's breast is the seal of the United States and in its talons are three arrows. The eagle and pentagon are surrounded by a silver circle consisting of thirteen stars and laurel and olive branches. At the top of the circle are five silver rays extending above the stars, which form the medal suspension. On the reverse of the pentagon is the raised inscription FROM THE SECRETARY OF DEFENSE TO with space to inscribe the recipient's name. On the reverse of the silver is the inscription FOR SUPERIOR SERVICE. The ribbon consists of a central stripe of red flanked on either side by stripes of white, blue and yellow. Additional awards of the Defense Superior Service Medal are denoted by oak leaf clusters.

LEGION OF MERIT

Instituted: 1942
Criteria: Exceptionally meritorious conduct in the performance of outstanding services to the United States
Devices: Bronze letter "V" (for valor), gold, silver star
Notes: Issued in four degrees (Legionnaire, Officer, Commander & Chief Commander) to foreign nationals

For exceptionally meritorious conduct in the performance of outstanding service. The Legion of Merit is worn after the Defense Superior Service Medal and before the Distinguished Flying Cross.

The Legion of Merit was established by an Act of Congress, which was approved 20 July 1942 and signed by President Franklin D. Roosevelt on 29 October 1942. The medal was instituted to fill the gap below the Distinguished Service Medal and provide an award that could be given for meritorious service in positions of considerable responsibility but below the positions of great responsibility called for by the criteria of the Distinguished Service Medal.

The Legion of Merit was designed by Colonel Townsend Heard and followed the basic design of the French Legion of Honor. The medal is a white enameled five-armed cross with ten gold tipped points. The cross is bordered in red enamel. In the center of the cross there is a blue enameled circle with thirteen stars surrounded by a border of gold clouds. Behind the cross is a gold circle bordered by a laurel wreath that is tied in a bow between the two lower arms of the cross. Gold crossed arrows appear between each of the arms of the cross. The same cross appears on the reverse of the medal except there is no enamel. In the center of the reverse is a rope circle for engraving the recipient's name. Contained in a second rope circle is the raised inscription ANNUIT COEPTIS MDCCLXXXII. The words ANNUIT COEPTIS are Latin meaning "God has favored our undertaking" and comes from the Great Seal of the United States. The MDCCLXXXII (1782) refers to the date George Washington founded the Badge of Military Merit, from which the Legion of Merit evolved. There is an outer

Continued on page 78

LEGION OF MERIT *continued from page 77*

ribbon circle with the raised inscription UNITED STATES OF AMERICA. The ribbon is ruby (pinkish-red) edged in white. Additional awards are denoted by five-sixteenth inch gold stars. A Combat Distinguishing Device (Bronze letter "V") may be authorized.

The Legion of Merit is awarded in four degrees to foreign nationals:

Chief Commander of the Legion of Merit, the highest degree, is intended for the heads of state of friendly foreign countries. The medal is in the form of a three inch diameter breast star. The service ribbon features a gold miniature of the medal set on a gold-colored bar engraved with two rows of arrow feathers on a ruby red ribbon edged in white.

Commander of the Legion of Merit, the second degree, is intended for high ranking officers of friendly foreign countries. The medal is suspended from a neck ribbon of the same color and design as the ribbon bar. The service ribbon features a silver miniature of the medal set on a silver-colored bar engraved with two rows of arrow feathers on a ruby red ribbon edged in white.

Officer of the Legion of Merit, the third degree, is intended for field grade officers of friendly foreign countries. The medal is identical to the Legionnaire grade with the addition of an eleven-sixteenth inch gold-colored miniature of the planchet (medal) attached to the suspension ribbon. The service ribbon has a five-sixteenth inch gold-colored miniature of the planchet (medal) on a ruby red ribbon edged in white.

Legionnaire of the Legion of Merit, the fourth degree is conferred upon friendly foreign service members and members of the Armed Forces of the United States. The Legion of Merit is commonly used to recognize field grade officers and Sergeants Major upon their retirement. The Legion of Merit is the second highest non combat award, ranking just after the Distinguished Service medals.

DISTINGUISHED FLYING CROSS

Gold

Silver

Gold

<u>Institued:</u> 1926
<u>Criteria:</u> Heroism or extraordinary achievement while participating in aerial flight
<u>Devices:</u> Bronze letter "V" (for valor), gold, silver star

For heroism or extraordinary achievement while participating in aerial flight. The Distinguished Flying Cross is worn after the Legion of Merit and before the Navy and Marine Corps Medal.

The Distinguished Flying Cross was established by an Act of Congress and approved in July 1926. The Act of Congress was implemented in January 1927 by President Coolidge. If the medal is awarded for an act of heroism, the act must involve voluntary action in the face of danger and be above the actions of others in a similar operation. If awarded for extraordinary achievement, it must have resulted in an accomplishment so outstanding or exceptional that the act clearly sets the individual apart from his or her comrades. This is specifically an aviation award.

The Distinguished Flying Cross was designed by the Army's Institute of Heraldry and is a bronze four bladed propeller surmounted on a bronze cross pattee. Behind the cross are bronze rays forming a square. The medal is suspended from a plain bronze suspender. The reverse of the medal is blank to allow for engraving the recipient's name. The ribbon is blue with a red center stripe bordered in white. The ribbon is outlined with a white stripe on each side. Additional awards are denoted by five-sixteenth inch diameter gold stars. The Combat Distinguishing Device (Bronze letter "V") may be authorized for qualifying service rendered after 4 April 1974.

NAVY AND MARINE CORPS MEDAL

Gold

Silver

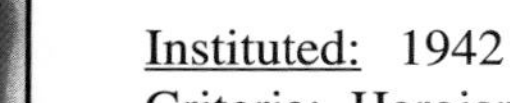

Instituted: 1942
Criteria: Heroism not involving actual conflict with an armed enemy of the United States
Devices: Gold, silver star
Notes: For acts of life-saving, action must be at great risk to one's own life

For heroism that involves the voluntary risk of life under conditions other than those of conflict with an opposing armed force. The Navy and Marine Corps Medal ranks after the Distinguished Flying Cross and before the Bronze Star Medal.

The Navy and Marine Corps Medal was established by an Act of Congress and approved on 7 August 1942. Its purpose was to recognize non-combat heroism. For acts of lifesaving, or attempted lifesaving, it is required that the action be performed at the risk of one's own life. The Navy and Marine Corps Medal is prized above many combat decorations by Marines who have received it.

The Navy and Marine Corps Medal was designed by Lt. Commander McClelland Barclay, USNR. The medal is a bronze octagon with an eagle perched upon a fouled anchor. Beneath the anchor is a globe and below that the inscription, HEROISM, in raised letters. The back of the medal is blank to allow for engraving the recipient's name. The ribbon consists of three equal strips of Navy blue, gold and scarlet. Additional awards are denoted by five-sixteenth inch gold stars.

BRONZE STAR MEDAL

Gold

Silver

Gold

Instituted: 1944
Criteria: Heroic or meritorious achievement or service not involving participation in aerial flight
Devices: Bronze letter "V" (for valor), gold, silver star

For heroic or meritorious achievement or service, not involving aerial flight in connection with operations against an opposing armed force. The Bronze Star Medal is worn after the Navy and Marine Corps Medal and before the Purple Heart.

The Bronze Star Medal was established by Executive Order and signed by President Franklin D. Roosevelt on 4 February 1944. The medal was instituted to provide ground forces with a medal comparable to the Air Medal and it was originally proposed as the "Ground Medal".

Like the Silver Star, the Bronze Star Medal was designed by the Atlanta retail jeweler - Bailey, Banks and Biddle. Its design reflects the intent that it be a companion to the Silver Star. The medal is a bronze five pointed star with a smaller raised and centered bronze star three-sixteenth inches in diameter. The reverse of the medal has the raised inscription HEROIC OR MERITORIOUS ACHIEVEMENT forming a circle in the center. The ribbon is red with a narrow blue center stripe. A thin white stripe borders the ribbon and the blue center stripe. Additional awards are denoted by five-sixteenth inch diameter gold stars. The combat distinguishing device (Bronze "V") may be authorized.

PURPLE HEART

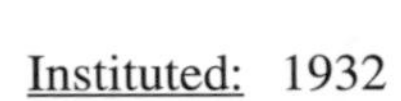

Instituted: 1932
Criteria: Awarded to any member of the U.S. Armed Forces killed or wounded in an armed conflict
Devices: Gold, silver star
Notes: Wound Ribbon appeared circa. 1917-18 but was never officially authorized

For wounds or death as a result of an act of any opposing armed force, as a result of an international terrorist attack or as a result of military operations while serving as a part of a peacekeeping force. The Purple Heart is worn after the Bronze Star Medal and before the Defense Meritorious Service Medal.

The Purple Heart stems from the Badge of Military Merit established by George Washington in 1782, which is the oldest American military decoration. Washington's Badge of Military Merit was referred to as the "Purple Heart" and was awarded for military merit. It is believed that the name comes from a wood called "purple heart", which is a smooth-grained plum-colored wood used with firearms and artillery. The wood was considered the best in the world for making gun carriages and mortar beds, because it could withstand extreme stress. The original medal was a heart-shaped purple cloth embroidered in silver with a wreath surrounding the word MERIT. It was designed by Pierre Charles L'Enfant in accordance with Washington's personal instructions. The Badge of Military Merit or "Purple Heart," though intended to be permanent, fell into disuse shortly after the Revolution and was all but forgotten as a military decoration.

The current Purple Heart Medal was revived by the War Department on 22 February 1932 at the urging of Army Chief of Staff General Douglas MacArthur, who was also the first recipient. The medal was authorized for the Navy and Marine Corps by the Department of the Navy on 21 January 1943. Although the Purple Heart was awarded for meritorious service between 1932 and 1943, the primary purpose was to recognize those who received wounds while in the service of the United States Military. With the development of awards such as the Legion of Merit the use of the Purple Heart came to be strictly limited to injuries sustained in combat. The current criteria states that it is to be awarded for wounds received while serving in any capacity with one of the U.S. Armed Forces after 5 April 1917. The wounds may have been received in combat against an enemy, while a member of a peacekeeping force, while a Prisoner of War, as a result of a terrorist attack, or as a result of a friendly fire incident in hostile territory. The 1996 Defense Authorization Act extended eligibility for the Purple Heart to prisoners of war before 25 April 1962; 1962 legislation had only authorized the medal to POWs after 25 April 1962. Wounds that qualify must have required treatment by a medical officer or must be a matter of official record.

The Purple Heart was designed by the Army's Institute of Heraldry from a design originally submitted by General Douglas MacArthur and modeled by John Sinnock, Chief Engraver at the Philadelphia Mint. The medal is a purple heart with a bronze gilt border and a bronze profile of George Washington in the center. Above the heart is a shield from George Washington's Coat of Arms between two sprays of green enameled leaves. On the reverse of the medal, below the Coat of Arms and leaves, there is a raised bronze heart with the raised inscription FOR MILITARY MERIT and room to inscribe the name of the recipient. Initially the medals were numbered, but this practice was discontinued in July 1943 as a cost-cutting measure. The ribbon is purple edged in white. Additional awards are denoted by five-sixteenth inch diameter gold stars.

1782 Badge of Military Merit

DEFENSE MERITORIOUS SERVICE MEDAL

Silver

Bronze

Instituted: 1977
Criteria: Noncombat meritorious achievement or service while assigned to a Joint Activity
Devices: Bronze, silver oak leaf cluster

Awarded in the name of the Secretary of Defense for recognition of non-combat meritorious achievement or exceptional service while serving in a joint activity. The Defense Meritorious Service Medal is worn after the Purple Heart and before the Meritorious Service Medal.

The Defense Meritorious Service Medal was established by Executive Order, which was signed by President Carter on 3 November 1977. The medal was instituted to recognize non-combat meritorious achievement or meritorious service by members of the Armed Forces assigned to the Office of the Secretary of Defense and other activities in the Department of Defense.

The Defense Meritorious Service Medal was designed by the Army's Institute of Heraldry and is similar in design to the Defense Distinguished Service Medal. The medal is bronze in color, featuring a pentagon superimposed by an eagle with outspread wings. The eagle and pentagon are surrounded by a circular wreath of laurel. The reverse is inscribed in raised letters, DEFENSE MERITORIOUS SERVICE set in three lines; around the bottom are the words UNITED STATES OF AMERICA. The ribbon is white with three blue center stripes and ruby-red border stripes. Additional awards of the Defense Meritorious Service Medal are denoted by oak leaf clusters.

MERITORIOUS SERVICE MEDAL

Silver

Gold

Instituted: 1969
Criteria: Outstanding noncombat meritorious achievement or service to the United States
Devices: Gold, silver star

For outstanding non-combat meritorious achievement or service to the United States. The Meritorious Service Medal is worn after the Defense Meritorious Service Medal and before the Air Medal.

The Meritorious Service Medal was established by Executive Order and signed by President Lyndon B. Johnson on 16 January 1969. The medal is considered a peacetime equivalent to the Bronze Star Medal.

The Meritorious Service Medal was designed by the Army's Institute of Heraldry. The medal is bronze with the upper part of the medal showing the upper half of a five-pointed star with six rays emanating outward.

The lower half of the medal shows an eagle with outstretched wings standing upon laurel branches forming the bottom of the medal. The reverse consists of the raised inscription UNITED STATES OF AMERICA around the top and MERITORIOUS SERVICE at the bottom; the center is blank allowing for the inscription of the recipient's name. The ribbon is ruby red with white border stripes. Additional awards are denoted by five-sixteenth inch diameter stars.

AIR MEDAL

Instituted: 1942
Criteria: Heroic actions or meritorious service while participating in aerial flight
Devices: Bronze letter "V," bronze numeral, gold numeral, bronze star, gold, silver star

For meritorious achievement while participating in aerial flight. The Air Medal may be awarded to individuals who, while serving in any capacity with the Armed Forces, distinguish themselves by heroism, outstanding achievement, or by meritorious service while participating in aerial flight, but to a lesser degree than which justify the award of the Distinguished Flying Cross. The Air Medal is worn after the Bronze Star Medal and before the Joint Service Commendation Medal. The Air Medal is considered by many to be the air version of the Bronze Star Medal.

The Air Medal was established by Executive Order, which was signed by President Franklin D. Roosevelt on 11 May 1942. The medal was intended to protect the prestige of the Distinguished Flying Cross and as a morale booster to recognize the same kind of acts that were recognized by the Distinguished Flying Cross, but to a lesser degree. The Navy and Marine Corps also use a system for awarding the medal for meritorious achievement while participating in sustained aerial flight operations based on the number of strikes or flights. Strikes are defined as sorties which encounter enemy opposition and flights are sorties without enemy opposition. The requirement calls for 10 strikes , 20 flights or 50 missions or 250 hours in direct combat support or any combination. The combination requires the accumulation of 20 points on the formula, a strike being valued at 2 points, a flight at 1 point , and a mission at .4 points. The Navy and Marine Corps distinguish between the award of the medal on a Strike/Flight basis and those awarded for Single Mission/Individual basis. This is done by placing a bronze arabic numeral (indicating the number of awards) on the ribbon bar on the wearer's left if the award is for Strike/ Flight. If the award is for individual heroism or achievement, a three-sixteenth inch bronze star is placed in the center of the ribbon for the first award, while five-sixteenth inch gold stars are used to denote additional individual awards (a silver star is used in lieu of five gold stars).

The use of stars to denote the number of Air Medals for Single Mission/Individual awards was discontinued during the period from 1 January 1980 to 22 November 1989 and a practice of using gold arabic numerals (indicating the number of awards) on the ribbon bar on the wearer's RIGHT was substituted. The current practice (since 22 November 1989) of denoting the number of Air Medals for Single Mission/Individual awards is with the use of five-sixteenth inch gold stars (a silver star is used in lieu of five gold stars).

The medal is a bronze compass rose of sixteen points. In the center of the compass is an American eagle swooping down in attack with a lightning bolt in each talon. The medal is suspended from the ribbon by a fleur-de-lis. The reverse of the medal has a blank raised disk to allow for inscribing the recipient's name. The ribbon is dark blue with orange stripes just inside each edge. A Combat Distinguishing Device "V" was authorized for use with the Air Medal effective 5 April 1974.

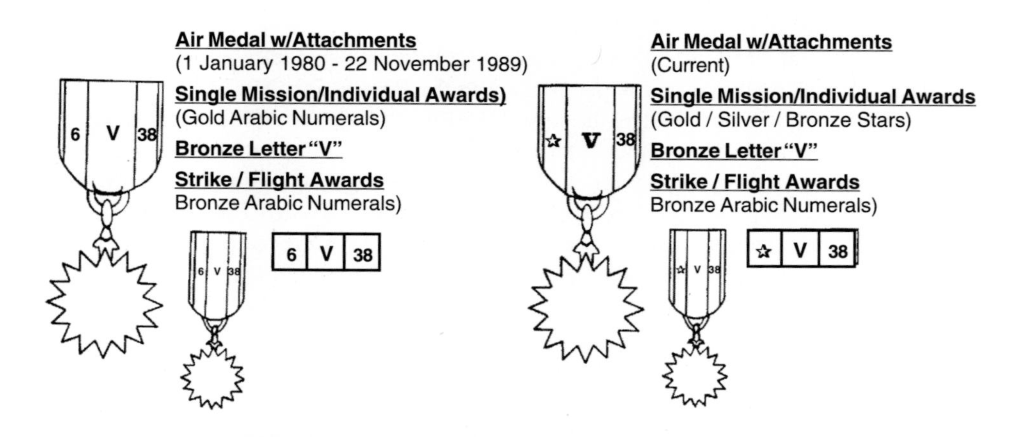

JOINT SERVICE COMMENDATION MEDAL

Bronze

Silver

Bronze

Instituted: 1963
Criteria: Meritorious service or achievement while assigned to a Joint Activity
Devices: Bronze letter "V" (for valor), bronze, silver oak leaf cluster

For meritorious achievement or service while assigned to a joint activity. The Joint Service Commendation Medal is worn after the Air Medal and before the Navy and Marine Corps Commendation Medal.

The Joint Service Commendation Medal was established by the Department of Defense on 25 June 1963 and made retroactive to 1 January 1963. The medal is awarded in the name of the Secretary of Defense to members of the Armed Forces who were assigned to the Office of the Secretary of Defense, the Joint Chiefs of Staff, the Defense Agencies or unified commands who have distinguished themselves by outstanding achievement or meritorious service, but to a lesser degree than required for the award of the Defense Meritorious Service Medal.

The Joint Service Commendation Medal was designed by the Army's Institute of Heraldry. The medal consists of four conjoined hexagons of green enamel edged in gold (two vertical and two horizontal). The upper hexagon contains thirteen gold stars and the lower hexagon has a gold heraldic delineation representing land, sea and air. In the center is a gold eagle taken from the Seal of the Department of Defense. The eagle and hexagons are surrounded by a circle of gold laurel leaves with gold bands. The reverse of the medal has a plaque for inscribing the recipient's name and the raised words FOR MILITARY (above the plaque) and MERIT (below the plaque). The ribbon consists of a green center stripe bordered on each side by with stripes of white, green, white and light blue. Additional awards are denoted by oak leaf clusters.

NAVY AND MARINE CORPS COMMENDATION MEDAL

Bronze

Silver

Gold

Instituted: 1950
Criteria: Meritorious service or achievement in a combat or noncombat situation based on sustained performance of a superlative nature
Devices: Bronze letter "V" (for valor), gold, silver star
Notes: Originally a ribbon-only award: "Secretary of the Navy Commendation for Achievement Award with Ribbon". Changed to present name in 1994.

For heroic and meritorious achievement or service. The Navy and Marine Corps Commendation Medal is worn after the Joint Service Commendation Medal and before the Joint Service Achievement Medal.

The Navy and Marine Corps Commendation Medal was originally established as a ribbon only award on 11 January 1944. The current medal was authorized by the Secretary of the Navy on 22 March 1950. The medal is awarded for both heroism and meritorious achievement. To be awarded for heroism, the act must be worthy of recognition, but to a lesser degree than required for the Bronze Star Medal in combat or the Navy and Marine Corps Medal in a non-combat situation. To be awarded for meritorious achievement, the act must be outstanding and worthy of special recognition, but to a lesser degree than required for the Bronze Star Medal in combat or the Meritorious Service Medal or Air Medal when in a non-combat situation.

The Navy and Marine Corps Commendation Medal was designed by the Army's Institute of Heraldry. The medal is a bronze hexagon with the eagle from the Seal of the Department of Defense in the center. The reverse of the medal has a plaque for inscribing the recipient's name and the raised words FOR MILITARY above the plaque and MERIT below the plaque. The ribbon is dark green with narrow a stripe of white near each edge. Additional awards of the Navy and Marine Corps Commendation Medal are denoted by five-sixteenth inch gold stars. A Combat Distinguishing Device (Bronze letter "V") may be authorized.

JOINT SERVICE ACHIEVEMENT MEDAL

Silver

Bronze

Institutéd: 1983
Criteria: Meritorious service or achievement while serving with a Joint Activity
Devices: Bronze, silver oak leaf cluster

For meritorious service or achievement while assigned to a joint activity after 3 August 1983. The Joint Service Achievement Medal is worn after the Navy and Marine Corps Commendation Medal and before the Navy and Marine Corps Achievement Medal.

The Joint Service Achievement Medal was established by the Department of Defense on 3 August 1983. The medal is awarded in the name of the Secretary of Defense to members of the Armed Forces (below the rank of colonel) assigned to the Office of the Secretary of Defense, the Joint Chiefs of Staff, the Defense Agencies, or unified commands who have distinguished themselves by outstanding achievement or meritorious service, but to a lesser degree than required for the award of the Joint Service Commendation Medal.

The Joint Service Achievement Medal was designed by the Army's Institute of Heraldry. The medal consists of a twelve pointed star with a gold eagle in the center. The eagle is taken from the Seal of the Department of Defense. The reverse of the medal has a circle composed of the raised words JOINT SERVICE ACHIEVEMENT AWARD. The ribbon has a thin red stripe in the center flanked on either side by a wide blue stripe, a thin white stripe, a narrow green stripe, a thin white stripe and a narrow dark blue border. Additional awards are denoted by oak leaf clusters.

NAVY AND MARINE CORPS ACHIEVEMENT MEDAL

Bronze

Silver

Gold

Instituted: 1961
Criteria: Meritorious service or achievement in a combat or noncombat situation based on sustained performance of a superlative nature
Devices: Bronze letter "V" (for valor), gold, silver star
Notes: Originally a ribbon-only award: "Secretary of the Navy Commendation for Achievement Award with Ribbon"

For junior officers and enlisted personnel whose professional and/or leadership achievements on or after 1 May 1961 are clearly of a superlative nature. The Navy and Marine Corps Achievement Medal is worn after the Joint Service Achievement Medal and before the Combat Action Ribbon.

The Navy and Marine Corps Achievement Medal was originally established as a ribbon only award on 1 May 1961. The current medal was authorized by the Secretary of the Navy on 17 July 1967. The medal is awarded for both professional and leadership achievement. To be awarded for professional achievement, the act must clearly exceed that which is normally required or expected, and it must be an important contribution to benefit the United States Naval Service. To be recognized for leadership achievement, the act must be noteworthy and contribute to the individual's unit mission.

The Navy and Marine Corps Achievement Medal was designed by the Army's Institute of Heraldry. The medal is a bronze square (with clipped corners) with a fouled anchor in the center. There is a star in each of the four corners. The reverse of the medal is blank to allow for engraving the recipient's name. The ribbon is myrtle green with stripes of orange near each edge. Additional awards of the Navy and Marine Corps Achievement Medal are denoted by five-sixteenth inch gold stars. A Combat Distinguishing Device (Bronze letter "V") may be authorized.

COMBAT ACTION RIBBON

Silver

Gold

Instituted: 1969
Criteria: Active participation in ground or air combat during specifically listed military operations
Devices: Gold, silver star
Notes: This is the only Navy personal decoration which has no associated medal (a "ribbon-only" award).

For active participation in specifically named ground or surface combat actions subsequent to 1 March 1961, while in the grade of Colonel or below. The Combat Action Ribbon is worn after the Navy and Marine Corps Achievement Medal and before the Presidential Unit Citation in a ribbon display. It is worn as the senior ribbon on the right breast when full-sized medals are worn on the left breast.

The Combat Action Ribbon was authorized by the Secretary of the Navy on 17 February 1969 and made retroactive to 6 December 1941. The principal requirement is that the individual was engaged in combat during which time he/she was under enemy fire and that his/her performance was satisfactory.

The Combat Action Ribbon is a ribbon only award. The ribbon is gold with thin center stripes of red, white and blue and border stripes of dark blue on the left and red on the right. Additional awards are authorized for each separate conflict/war and are represented by five-sixteenth inch gold stars.

PRESIDENTIAL UNIT CITATION

Gold

Silver

Bronze

Instituted: 1942
Criteria: Awarded to Navy/Marine Corps units for extraordinary heroism in action against an armed enemy.
Devices: Bronze, silver star, gold "N" (USS Nautilus), gold globe (USS Triton)

For service in a unit named by the President for outstanding performance in action. The Presidential Unit Citation is worn after the Combat Action Ribbon and before the Joint Meritorious Unit Award.

The Presidential Unit Citation was established by Executive Order on 6 February 1942 and amended on 28 June 1943. It is awarded by the Secretary of the Navy in the name of the President. The citation is conferred on units for displaying extraordinary heroism; effective 16 October 1941. The degree of heroism required is the same as that which is required for the award of the Navy Cross to an individual. Only an individual assigned to the unit when the award was granted may wear the ribbon on the uniform.

The Presidential Unit Citation is a ribbon only award. The ribbon consists of three equal horizontal stripes of navy blue (top), gold (middle) and red (bottom). Additional awards of the Presidential Unit Citation are denoted by three-sixteenth inch bronze stars.

JOINT MERITORIOUS UNIT AWARD

Silver

Bronze

Instituted: 1981
Criteria: Awarded to Joint Service units for meritorious achievement or service in combat or extreme circumstances
Devices: Bronze, silver oak leaf cluster

Recognizes joint units or activities for meritorious achievement or service superior to that which is normally expected. The Joint Meritorious Unit Award is worn after the Presidential Unit Citation and before the Navy Unit Commendation.

The Joint Meritorious Award was authorized by the Secretary of Defense on 10 June 1981 and was originally called the Department of Defense Meritorious Award. It is awarded in the name of the Secretary of Defense for meritorious service, superior to that which would normally be expected during combat, or declared national emergency, or under extraordinary circumstances that involve national interest. The service performed by the unit would be similar to that performed by an individual awarded the Defense Superior Service Medal. The award is retroactive to 23 January 1979.

The Joint Meritorious Award is a ribbon only award. The ribbon is similar to the Defense Superior Service Medal ribbon with a gold metal frame with laurel leaves. Like the Defense Superior Service Medal, the ribbon consists of a central stripe of red flanked on either side by stripes of white, blue and yellow with blue edges. Additional awards are denoted by oak leaf clusters.

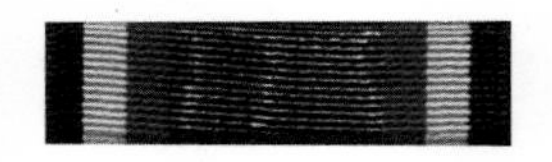

NAVY UNIT COMMENDATION

Instituted: 1944
Criteria: Awarded to units Navy/Marine Corps for outstanding heroism in action or extremely meritorious service
Devices: Bronze, silver star

For outstanding heroism in action or extremely meritorious service not involving combat, but in support of military operations. The Navy Unit Commendation is worn after the Joint Meritorious Unit Award and before the Navy Meritorious Unit Commendation.

The Navy Unit Commendation was established by the Secretary of the Navy on 18 December 1944. The Commendation is awarded by the Secretary of the Navy with the approval of the President. The Commendation is made to units, which, subsequent to 6 December 1941, distinguish themselves by outstanding heroism in action against an enemy, but to a lesser degree than required for the Presidential Unit Citation. The Commendation may also be awarded for extremely meritorious service not involving combat, but in support of military operations, which is outstanding when compared to other units performing similar service.

The Navy Unit Commendation is a ribbon only award. The ribbon is dark green with narrow border stripes of red, gold and blue. Additional awards are denoted by three-sixteenth inch bronze and silver stars.

NAVY MERITORIOUS UNIT COMMENDATION RIBBON

Instituted: 1967
Criteria: Awarded to Navy/Marine Corps units for valorous actions or meritorious achievement (combat or noncombat)
Devices: Bronze, silver star

Silver

For any unit which distinguishes itself by valorous or meritorious achievement or service or outstanding service. The Meritorious Unit Commendation is worn after the Navy Unit Commendation and before the Navy "E" ribbon.

The Meritorious Unit Commendation was established by the Secretary of the Navy on 17 July 1967. The Commendation is awarded by the Secretary of the Navy to units which distinguish themselves by either valorous or meritorious achievement considered outstanding, but to a lesser degree than required for the Navy Unit Commendation. The Commendation may be awarded for services in combat or non-combat situations.

The Navy Meritorious Unit Commendation is a ribbon only award. The ribbon is dark green with a narrow red center stripe flanked on either side by stripes of gold, navy blue and gold. Additional awards are denoted by three-sixteenth inch bronze and silver stars.

NAVY "E" RIBBON

Instituted: 1976
Criteria: Awarded to ships or squadrons which have won battle efficiency competitions
Devices: Silver letter "E", wreathed silver letter "E"

To recognize individuals who were permanently assigned to ships or squadrons that won the battle efficiency competitions subsequent to 1 July 1974. The Navy "E" Ribbon is worn after the Meritorious Unit Commendation and before the Prisoner of War Medal on a ribbon display and on the right breast before the Sea Service Deployment Ribbon if wearing full-sized medals.

The Navy "E" Ribbon was established in June 1976 and is authorized to be worn by all personnel who served as permanent members of ship's company or squadrons winning the Battle Efficiency Award.

The Navy "E" Ribbon is a ribbon only award. The ribbon is navy blue with borders of white and gold with a silver "E" in the center. Additional awards are denoted by additional "E"s. The fourth (final) award is denoted by an "E" surrounded by a silver wreath.

PRISONER OF WAR MEDAL

Silver

Bronze

Institued: 1985, retroactive to 5 April 1917
Criteria: Awarded to any member of the U.S. Armed Forces taken prisoner during any armed conflict dating from World War I
Devices: Bronze, silver star

Awarded to any person who was taken prisoner of war and held captive after 5 April 1917. The Prisoner of War Medal is worn after the Navy "E" ribbon and before the Good Conduct Medal.

The Prisoner of War Medal was authorized by Public Law Number 99-145 in 1985 and may be awarded to any person who was taken prisoner or held captive while engaged in an action against an enemy of the United States; while engaged in military operations involving conflict with an opposing armed force; or while serving with friendly forces engaged in armed conflict against an opposing armed force in which the United States is not a belligerent party. The recipient's conduct, while a prisoner, must have been honorable.

The Prisoner of War Medal was designed by the Army's Institute of Heraldry. The medal is a bronze disc with an American eagle centered and completely surrounded by a ring of barbed wire and bayonet points. The reverse of the medal has a raised inscription AWARDED TO (with a space for the recipient's name) and FOR HONORABLE SERVICE WHILE A PRISONER OF WAR set in three lines. Below this is the shield of the United States and the words UNITED STATES OF AMERICA. The ribbon is black with thin border stripes of white, blue, white and red. Additional awards are denoted by three-sixteenth inch bronze stars.

NAVY GOOD CONDUCT MEDAL

Silver

Bronze

Instituted: 1884
Criteria: Outstanding performance and conduct during 3 years of continuous active enlisted service in the U.S Navy
Devices: Bronze, silver star

Based on good conduct and faithful service for three year periods of continuous active service. The Navy Good Conduct Medal is worn after the Prisoner of War Medal and before the Naval Reserve Meritorious Service Medal.

The Navy Good Conduct Medal was authorized on 21 November 1884. The medal is awarded to enlisted personnel of the United States Navy and Naval Reserve (active duty) for creditable above average professional performance, military behavior, leadership, military appearance and adaptability. Those receiving the award must have had no convictions by court martial and no nonjudicial punishment during that three year period (there was a time from November 1963 to January 1996 when the period was four years). For the first award, the medal may be awarded to the next of kin in those cases where the individual is missing in action or dies of wounds received in combat. Naval personnel may also receive the medal if separated from the service as a result of wounds incurred in combat.

The forerunner of the Navy Good Conduct Medal was the Navy Good Conduct Badge which was established in 1868 by the Secretary of the Navy, making it our Country's second oldest award. The badge, in use from 1868 to 1884, was awarded to men holding a Continuous Service Certificate awarded upon the successful completion of a term of enlistment. In these early days, any seaman who qualified for three awards was promoted to petty officer.

Continued on page 88

NAVY GOOD CONDUCT MEDAL

Continued from page 87

The Good Conduct Badge was a Maltese Cross with a circular medallion in the center. The medallion was bordered with a border inscribed around the edge with the words FIDELITY - ZEAL - OBEDIENCE and U.S.A. in the center. The cross was suspended from a 1/2 inch wide red, white and blue ribbon .

In 1880 the Navy redesigned the Good Conduct Badge. The new medallion was proposed by Commodore Winfield Scott Schley from the design used on the letterhead of the Navy Department's Bureau of Equipment and Recruiting. This new medallion was suspended from a 1-5/8 inch wide red ribbon with thin border stripes of white and blue. In 1884 the medal was redesigned and in 1896 the award period was changed to three years of continuous active service. This new medal maintained the 1880 design and was suspended from a maroon ribbon by a straight bar clasp. Subsequent awards were recognized by the addition of clasps, which were placed on the suspension ribbon between the top of the ribbon and the medallion. These clasps were bordered with rope and were engraved with ship or duty station of the recipient. During World War I, medals were impressed with rim numbers and many were issued without engraving. In the 1930's the practice of engraving the ship or duty station on the clasps was replaced by the enlistment discharge date. In 1942 engraving clasps was discontinued, and the clasps showed SECOND AWARD, THIRD AWARD, etc. in raised letters. Following World War II, the Navy changed its practice of engraving to stamping the recipient's information on the medal's reverse and the current practice of using three-sixteenth bronze stars was adopted to denote additional awards.

The current medal is a circular bronze disc with a raised anchor and anchor chain circling a depiction of the U. S. S. Constitution and the words CONSTITUTION and the words UNITED STATES NAVY. The reverse side of the medal has the raised inscription FIDELITY - ZEAL - OBEDIENCE around the border with room in the center to stamp the recipient's name.

The ribbon of the current Navy Good Conduct Medal is maroon. Additional awards are denoted by three-sixteenth bronze stars.

Good Conduct Badge 1869-1884

The first good conduct recognition was the award of the Good Conduct Badge that was awarded from 1869 to 1884.

The early version of the Good Conduct Medal was suspended permanently from a bar and had the name of the ship or duty station engraved on rope-bordered bars. This versionof the medal also was engraved with the recipient's Continuous Service number, and dates of service.

Early Good Conduct Medal

RESERVE SPECIAL COMMENDATION RIBBON

Instituted: 1946
Criteria: Awarded to Reserve Officers with 4 years of successful command and a total Reserve service of 10 years.
Devices: None

For those officers of the Organized Reserve who had officially commanded in a meritorious manner for a period of four years between 1 January 1930 and 7 December 1941, an organized battalion, squadron, or separate division (not part of a battalion) of the Naval Reserve, or an organized battalion or squadron of the Marine Corps Reserve, and have had a total service in the Reserve of not less than ten years. The four year period need not have been continuous, but officers must have been regularly assigned to command such units for a total of four years.

The Reserve Special Commendation Ribbon was authorized by the Secretary of the Navy on 16 April 1946, but owing to the applicable dates, was obsolete on the day it was instituted.

The Reserve Special Commendation Ribbon was a ribbon only award. The ribbon is red with a wide green center stripe and thin border stripes of gold and blue.

NAVAL RESERVE MERITORIOUS SERVICE MEDAL

Instituted: 1964
Criteria: Outstanding performance and conduct during four years of enlisted service in the Naval Reserve
Devices: Bronze star

For naval reservists who fulfill, with distinction, obligations above a level that is normally expected. The Naval Reserve Meritorious Service Medal is worn after the Navy Good Conduct Medal (or Reserve Special Commendation Ribbon, before it became obsolete) and before the Fleet Marine Force Ribbon.

The Naval Reserve Meritorious Service Medal was authorized on 12 September 1959 originally as a ribbon only award. The medal was authorized on 22 June 1964 with eligibility back-dated to 1 July 1958. The award is made on a selected basis to U. S. Navy Reservists who fulfill, with distinction, the obligations of an inactive reservist at a higher level than normally expected. The obligations pertain to attendance and performance.

The Naval Reserve Meritorious Service Medal is a bronze disc showing a fouled anchor covered with a scroll with the raised words MERITORIOUS SERVICE. The words UNITED STATES NAVAL RESERVE circle the anchor. The reverse of the medal is blank. The ribbon is red with a blue center stripe and thin border stripes of gold and blue. Additional awards are denoted by three-sixteenth inch bronze stars.

FLEET MARINE FORCE RIBBON

Instituted: 1984
Criteria: Active participation by professionally skilled Navy personnel with the Fleet Marine Force
Devices: None

To recognize active service of Naval personnel who serve with the Fleet Marine Force. The Navy Fleet Marine Force Ribbon is worn after the Naval Reserve Meritorious Service Medal and before the Navy Expeditionary Medal.

The Navy Fleet Marine Force Ribbon was authorized by the Secretary of the Navy on 1 September 1984. The ribbon recognizes officers and enlisted personnel of the U. S. Navy who have served a minimum of 12 months with the Fleet Marine Force of the U. S. Marine Corps and demonstrated exceptional Navy qualification in providing support in a combat environment. Personnel must satisfactorily complete the Marine Corps Essential Subjects Test and Physical Fitness Test.

The Navy Fleet Marine Force Ribbon is a ribbon only award. The ribbon consists of a wide center stripe of Marine Corps red bordered by wide Navy blue stripes. The wide stripes are separated by thin gold stripes and the ribbon is edged in light blue.

NAVY EXPEDITIONARY MEDAL

Silver

Bronze

Instituted: 1936
Dates: 1936-Present
Criteria: Landings on foreign territory and operations against armed opposition for which no specific campaign medal has been authorized
Devices: Silver letter "W," bronze, silver star
Bar: "Wake Island"

Wake Island Bar

For opposed landing on a foreign territory or operations deserving special recognition. The Navy Expeditionary Medal is worn after the Navy Fleet Marine Force Ribbon and before the China Service Medal.

The Navy Expeditionary Medal was authorized on 5 August 1936. The medal is awarded to members of the Marine Corps who have engaged in operations against armed opposition in foreign territory, or have served in situations warranting special recognition where no other campaign medal was awarded. Many operations have qualified for the award (eight since World War II), the latest of which being operations in the Persian Gulf prior to Desert Shield/Desert Storm.

The Navy Expeditionary Medal is a bronze disc showing a sailor beaching a boat containing an officer and Marines with a flag of the United States and the word EXPEDITIONS. The reverse of the medal shows an American eagle perched on an anchor and laurel branches. On either side of the eagle are the words FOR SERVICE. Above, in a semicircle is a raised inscription UNITED STATES NAVY. The ribbon is Navy blue and gold, the official colors of the Navy. The ribbon has a wide blue center stripe flanked by gold with narrow blue edges. Additional awards are denoted by three-sixteenth inch bronze stars. For those who served in defense of Wake Island there is a one-quarter inch silver "W" for the ribbon bar and a clasp for the medal inscribed WAKE ISLAND.

CHINA SERVICE MEDAL

Bronze

Instituted: 1940
Dates: 1937-39, 1945-57
Criteria: Service ashore in china or on-board naval vessels during either of the above periods
Devices: Bronze star
Notes: Medal was reinstituted in 1947 for extended service during dates shown above

For service in China, during the periods just prior to and just following World War II. The China Service Medal is worn after the Marine Corps Expeditionary Medal and Before the American Defense Service Medal.

The China Service Medal was authorized by the Department of the Navy on 23 August 1940 for members of the Navy and Marine Corps who served in China or were attached to ships in the area during 7 July 1937 to 7 September 1939. The second period was for those who were present for duty during operations in China, Taiwan, and the Matsu Straits during the 2 September 1947 to 1 April 1957.

The China Service Medal was designed by George Snowden. The medal is a bronze disc showing a Chinese junk under full sail with the raised inscribed words CHINA above and SERVICE below. The reverse of the medal shows an American eagle perched on an anchor and laurel branches. On either side of the eagle are the words FOR SERVICE. Above, in a semicircle is a raised inscription UNITED STATES NAVY. The ribbon is yellow with a narrow red stripe near each edge. If an individual served during both periods, a bronze three-sixteenth inch star is worn.

AMERICAN DEFENSE SERVICE MEDAL

Bronze

Bronze

Instituted: 1941
Dates: 1939-41
Criteria: For active duty during national and limited emergencies just prior to World War II
Devices: Bronze star, bronze letter "A" (not worn with bronze star)
Bars: "Base," and "Fleet"

For service during the period of national emergency just prior to World War II. The American Defense Service Medal is worn after the China Service Medal and before the American Campaign Medal.

The American Defense Service Medal was established by Executive Order on 28 June 1941. The Department of the Navy authorized the medal for the Navy and Marine Corps on 20 April 1942. The medal was awarded to members of the Armed Forces for service during the period of 8 September 1939 to 7 December 1941.

The American Defense Service Medal was designed by Lee Lawrie. The medal is a circular bronze disc with a female figure (Liberty) brandishing a sword and holding a shield. The figure is standing on an oak branch with four leaves representing four services. Above are the inscribed words AMERICAN DEFENSE. The reverse has the raised inscription FOR SERVICE DURING THE LIMITED EMERGENCY PROCLAIMED BY THE PRESIDENT ON 8 SEPTEMBER 1939 OR DURING THE LIMITED EMERGENCY PROCLAIMED BY THE PRESIDENT ON 27 MAY 1941 set in twelve lines. Below this is a spray of seven oak leaves. The ribbon is yellow with narrow red, white and blue stripes near each edge. The Navy and Marine Corps had two clasps for the medal. The FLEET clasp was awarded for service with the fleet on the high seas and the BASE clasp was for service at bases outside the United States. A three-sixteenth inch bronze star was worn on the ribbon bar to denote the award of a clasp. In addition to the clasps, the block letter "A" was authorized for wear on the ribbon bar and medal suspension ribbon by personnel who served in the Atlantic Fleet on the high seas prior to the outbreak of World War II.

AMERICAN CAMPAIGN MEDAL

Bronze

Instituted: 1942
Dates: 1941-46
Criteria: Service outside the U.S. in the American theater for 30 days, or within the continental U.S. for one year.
Devices: Bronze, silver star, bronze Marine Corps device

For service during World War II within the American Theater of Operations. The American Campaign Medal is worn after the American Defense Service Medal and before the European - African - Middle Eastern Campaign Medal.

The American Campaign Medal was established by Executive Order on 6 November 1942 and amended on 15 March 1946, which established a closing date. The medal is awarded to all members of the Armed Forces who served in the American Theater of Operations during the period from 7 December 1941 to 2 March 1946 or were awarded a combat decoration while in combat against the enemy. The service must have been an aggregate of one year within the continental United States, or thirty consecutive days outside the continental United States, or sixty nonconsecutive days outside the continental United States, but within the American Theater of Operations. Maps of the three theaters of operation during World War II were drawn on 6 November 1942 to define the American Theater, the European - African - Middle Eastern Theater and the Asiatic - Pacific Theater.

The American Campaign Medal was designed by the Army's Institute of Heraldry. The medal is a circular bronze disc showing a Navy cruiser, a B-24 bomber and a sinking enemy submarine above three waves. Also shown in the background are some buildings representing the United States. Above is the raised inscription AMERICAN CAMPAIGN. The reverse of the medal shows an American eagle standing on a rock. On the left of the eagle are the raised inscribed dates 1941 - 1945 and on the right UNITED STATES OF AMERICA. The ribbon is azure blue with three narrow stripes of red, white and blue (United States) in the center and four stripes of white, red (Japan), black and white (Germany) near the edges. Three-sixteenth inch bronze stars indicated participation in specialized anti-submarine, escort or special operations. Nine special operations warrant battle stars.

EUROPEAN-AFRICAN-MIDDLE EASTERN CAMPAIGN MEDAL

Silver

Bronze

Instituted: 1942
Dates: 1941-45
Criteria: Service in the European-African-Middle Eastern theater for 30 days or receipt of any combat decoration
Devices: Bronze, silver star, bronze Marine Corps device

For service during World War II within the European, African, Middle Eastern Theater of Operations. The European - African - Middle Eastern Campaign Medal is worn after the American Campaign Medal and before the Asiatic - Pacific Campaign Medal.

The European - African - Middle Eastern Campaign Medal was established by Executive Order on 6 November 1942 and amended on 15 March 1946, which established a closing date. The medal is awarded to all members of the Armed Forces who served in the European, African, Middle Eastern Theater of Operations during the period from 7 December 1941 to 2 March 1946. The service must have been as a member of the Armed Forces on permanent assignment in the theater, or within the theater on temporary assignment for thirty consecutive days, or sixty nonconsecutive days, or the award of a combat decoration in the theater. Maps of the three theaters of operations during World War II were drawn on 6 November 1942 to define the American Theater, the European - African - Middle Eastern Theater and the Asiatic - Pacific Theater.

The European - African - Middle Eastern Campaign Medal was designed by the Army's Institute of Heraldry. The medal is a bronze disc showing troops assaulting a beach. A LST (Landing Ship Tank) and an airplane are in the background. Above is the raised inscription EUROPEAN AFRICAN MIDDLE EASTERN CAMPAIGN set in three lines. The reverse of the medal shows an American eagle standing on a rock. On the left of the eagle are the raised inscribed dates 1941 - 1945 and on the right UNITED STATES OF AMERICA. The ribbon has narrow center stripes of red, white and blue (United States). These are flanked by wide stripes of green on the left by narrow stripes of green, white and red (Italy), and on the right by narrow stripes of white, black and white (Germany). The stripes at the edges are brown (Africa). Participation in specific combat operations is denoted by three-sixteenth inch bronze stars. A three-sixteenth inch silver star is worn in lieu of five bronze stars. Naval personnel who were attached to units of the Fleet Marine Force (FMF) during this campaign are entitled to wear a small Marine Corps emblem device on the ribbon.

The nine Navy campaign designations for the European - African Middle Eastern Campaign Medal are:

North African Occupation,
8 November 1942 -9 July 1943

Sicilian Occupation,
9-15 July and 28 July -17 August 1943

Salerno Landings,
9-12 September 1943

West Coast of Italy Operations,
22 January-17 June 1944

Invasion of Normandy,
6-25 June 1944

Northwest Greenland Operation,
10 June-17 November 1944

Invasion of Southern France,
15 August-25 September 1944

Reinforcement of Malta,
14-21 April and 3-16 May 1942

Escort, Antisubmarine, Armed Guard and Special Operations,
16 December 1941-27 February 1943

ASIATIC - PACIFIC CAMPAIGN MEDAL

Silver

Bronze

Instituted: 1942
Dates: 1941-46
Criteria: Service in the Asiatic-Pacific theater for 30 day or receipt of any combat decoration
Devices: Bronze, silver star, bronze Marine Corps device

For service during World War II within the Asiatic Pacific Theater of Operations. The Asiatic - Pacific Campaign Medal is worn after the European - African - Middle Eastern Campaign Medal and before the World War II Victory Medal.

The Asiatic - Pacific Campaign Medal was established by Executive Order on 6 November 1942 and amended on 15 March 1946, which established a closing date. The medal is awarded to all members of the Armed Forces who served in the Asiatic - Pacific Theater of Operations during the period from 7 December 1941 to 2 March 1946. The service must have been as a member of the Armed Forces on permanent assignment in the theater, or within the theater on temporary assignment for thirty consecutive days, or sixty nonconsecutive days, or the award of a combat decoration in the theater. Maps of the three theaters of operations during World War II were drawn on 6 November 1942 to define the American Theater, the European - African - Middle Eastern Theater and the Asiatic - Pacific Theater.

The Asiatic - Pacific Campaign Medal was designed by the Army's Institute of Heraldry. The medal is a bronze disc showing troops landing in a tropical setting with a palm tree, battleship, aircraft carrier and submarine in the background. At the top of the medal, around the edge, are the words ASIATIC PACIFIC CAMPAIGN. The reverse of the medal shows an American eagle standing on a rock. On the left of the eagle are the raised inscribed dates 1941 - 1945 and on the right UNITED STATES OF AMERICA. The ribbon is yellow-orange with narrow center stripes of red, white and blue (United States). Near the edges are narrow white, red and white stripes (Japan). Participation in specific combat operations is denoted by three-sixteenth inch bronze stars. A three-sixteenth inch silver star is worn in lieu of five bronze stars. Naval personnel who were attached to units of the Fleet Marine Force (FMF) during this campaign are entitled to wear a small Marine Corps emblem device on the ribbon.

The forty-four Navy campaign designations for the Asiatic - Pacific Campaign Medal are:

Pearl Harbor-Midway, 1941
Wake Island, 1941
Philippine Islands Operation, 1941-1942
Netherlands East Indies, 1941-1942
Pacific Raids, 1942
Coral Sea, 1942
Midway, 1942
Guadalcanal, Tulagi Landings, 1942
Capture and Defense of Guadalcanal, 1942-1943
Makin Raid, 1942
Eastern Solomons (Stewart Isl.), 1942
Buin-Faisi-Tonolai, 1942
Cape Esperance (Second Savo), 1942
Santa Cruz Islands, 1942
Guadalcanal (Third Savo), 1942
Tassafaronga (Fourth Savo), 1942
Eastern New Guinea, 1942-1944
Rennel Island Operation, 1943
Solomon Islands Consolidation, 1943-1945
Aleutians Operations, 1943
New Georgia Group Operation, 1943
Bismarck Archipelago, 1943-1944
Pacific Raids, 1943
Treasury-Bougainville Operation, 1943
Gilbert Island Operation, 1943
Marshall Islands Operation, 1943-1944
Asiatic-Pacific Raids, 1944
Western New Guinea, 1944-1945
Marianas Operation
Hollandia Operation, 1944
Western Caroline Islands, 1944
Leyte Operation, 1944
Luzon Operation, 1944-1945
Iwo Jima Operation, 1945
Okinawa Gunto Operation, 1945
Third Fleet Operations against Japan, 1945
Kurile Islands Operation, 1944-1945
Borneo Operation, 1945
Tinian Capture and Occupation, 1945
Consolidation of Southern Philippines, 1945
Manila Bay-Bicol Operation, 1945
Escort, Antisubmarine, Armed Guard and Special Operations
Submarine War Patrols (Pacific)
Minesweeping Operations Pacific, 1945-1946

WORLD WAR II VICTORY MEDAL

Instituted: 1945
Dates: 1941-46
Criteria: Awarded for service in the U.S. Armed Forces during the above period
Devices: None

For service during World War II. The World War II Victory Medal is worn after the European - African - Middle Eastern Campaign Medal and before the Navy Occupation Service Medal.

The World War II Victory Medal was authorized on 6 July 1945. The medal was awarded to all members of the Armed Forces who served on active duty during the period from 7 December 1941 to 31 December 1946. It was also awarded to members of the Philippine Armed Forces.

The World War II Victory Medal was designed by the Army's Institute of Heraldry. The medal is a bronze disc showing the figure of Liberation holding the hilt of a broken sword in her right hand and the broken blade in her left; her right foot is resting on an ancient war helmet. Also at the figure's feet, behind the helmet, is a sun with rays spreading upward. The figure separates the raised inscription WORLD WAR II. The reverse of the medal has raised inscriptions FREEDOM FROM FEAR AND WANT and FREEDOM OF SPEECH AND RELIGION, separated by a palm branch. Around the edge of the reverse are the inscribed words UNITED STATES OF AMERICA and the dates 1941 - 1945. The ribbon has a wide dark red center stripe bordered by narrow stripes of white. The borders consist of bands of color starting in the center with red flanked by orange, yellow, green, blue and navy. No attachments are authorized for the World War II Victory Medal, although some veterans state that they received the medal with a three-sixteenth inch bronze star affixed. They believe that the star was to distinguish them as having been overseas at the end of the war. Although this may have been the practice, I have found no documentation to support this claim.

NAVY OCCUPATION SERVICE MEDAL

Gold Airplane

Instituted: 1948
Dates: 1945-55 (Berlin: 1945-90)
Criteria: 30 consecutive days of service in occupied territories of former enemies during above period
Devices: Gold airplane
Bars: "Europe," "Asia"

EUROPE

ASIA

For thirty consecutive days of service in occupied zones following World War II. The Navy Occupation Service Medal is worn after the World War II Victory Medal and before the Medal for Humane Action.

The Navy Occupation Service Medal was authorized by ALNAV 24 on 22 January 1947 and Navy Department GO on 28 January 1948. The medal was awarded for occupation duty in Japan and Korea from 2 September 1945 to 27 April 1952. The medal was also awarded for occupation service in Germany, Italy, Trieste and Austria although Naval personnel were not involved in these occupied zones.

The Navy Occupation Service Medal was designed by the Army's Institute of Heraldry. The medal is a bronze disc showing Neptune, god of the sea, riding a sea serpent with the head and front legs of a horse. Neptune is holding a trident in his right hand and pointing to an image of land, at the left of the medal, with his left hand. The lower front of the medal depicts the ocean with the words OCCUPATION SERVICE in two lines. The reverse of the medal shows an American eagle perched on an anchor and laurel branches. On either side of the eagle are the words FOR SERVICE. Above, in a semicircle is a raised inscription UNITED STATES NAVY. The ribbon has two wide stripes of red and black in the center with border stripes of white. Clasps, similar to those used on the World War I Victory Medal, are used to denote service in "EUROPE" and "ASIA", which are authorized for wear with the medal. There are no devices authorized for wear on the ribbon bar which represents these clasps. In addition, Navy and Marine personnel who served 90 consecutive days in support of the Berlin Airlift (1948-1949) are authorized to wear the Berlin Airlift device, a three-eighths inch gold C-54 airplane, on the ribbon bar and suspension ribbon.

MEDAL FOR HUMANE ACTION

Instituted: 1949
Dates: 1948-49
Criteria: 120 consecutive days of service participating in the Berlin Airlift or in support thereof
Devices: None

For service in support of the Berlin Airlift. The Medal for Humane Action is worn after the Navy Occupation Service Medal and before the National Defense Service Medal.

The Medal for Humane Action was authorized by Congress on 20 July 1949 for service of at least 120 days while participating in , or providing direct support for, the Berlin Airlift. The airlift period was from 26 June 1948 to 30 September, 1949.

The Medal for Humane Action was designed by the Army's Institute of Heraldry. The medal is a bronze disc with a depiction of a C-54 aircraft, which carried the majority of the airlift. The C-54 is centered on the medal above the Berlin coat of arms, which lies in the center of a wreath of wheat. The reverse has the American eagle, from the seal of the Department of Defense, above the raised inscription TO SUPPLY NECESSITIES OF LIFE TO THE PEOPLE OF BERLIN GERMANY. The words FOR HUMANE ACTION appear arched above the eagle. The ribbon is medium blue with borders of black edged (inboard) by narrow white stripes. In the center are narrow stripes of white, red and white. No attachments are authorized for this medal.

NATIONAL DEFENSE SERVICE MEDAL

★ Bronze

Instituted: 1953
Dates: 1950-54, 1961-74, 1990-95
Criteria: Any honorable active duty service during any of the above periods
Devices: Bronze star
Notes: Reinstituted in 1966 and 1991 for Vietnam and Southwest Asia (Gulf War) actions respectively

For active federal service in the Armed Forces 1950-54, 1961-74 and 1990-95. The National Defense Service Medal is worn after the Medal for Humane Action and before the Korean Service Medal.

The National Defense Service Medal was authorized by Executive Order on 22 April 1953. The medal was awarded for active service during the Korean War (27 June 1950 to 27 July 1954), the Vietnam War (1 January 1961 to 14 August 1974) and Desert Shield/Desert Storm (2 August 1990 to 30 November 1995).

The National Defense Service Medal was designed by the Army's Institute of Heraldry. The medal is a bronze disc with the American bald eagle perched on a sword and palm branch. Above the eagle, in a semicircle, are the raised engraved words NATIONAL DEFENSE. The reverse shows a shield from the Great Seal of the United States with an oak and laurel leave spray in a semicircle below it. The ribbon is red with a wide center stripe of yellow bordered by thin stripes of red, white, blue and white. Three-sixteenth inch bronze stars are used to denote each additional period of qualifying service.

KOREAN SERVICE MEDAL

Silver

Bronze

Instituted: 1950
Dates: 1950-54
Criteria: Participation in military operations within the Korean area during the above period
Devices: Bronze, silver star, bronze Marine Corps device

For participation in operations in the Korean area during the Korean War. The Korean Service Medal is worn after the National Defense Service Medal and before the Antarctic Service Medal.

The Korean Service Medal was authorized by Executive Order on 8 November, 1950. The medal was awarded to members of the Armed Forces for service in Korea from 27 June 1950 to 27 July 1954. The service required was one or more days in the designated area while attached to or serving with an organization or on a naval vessel that was participating in combat operations or in direct support of combat missions. Thirty consecutive days or sixty nonconsecutive days were required for individuals on Temporary Additional Duty unless the personnel participated in actual combat, in which case the time period was waived.

The Korean Service Medal was designed by the Army's Institute of Heraldry. The medal is a circular bronze disc showing a Korean gateway encircled by a raised inscription KOREAN SERVICE. The reverse of the medal shows the symbol from the Korean national flag representing the unity of all being. Around this symbol is the raised inscription UNITED STATES OF AMERICA with an oak and laurel spray at the bottom. The ribbon is light blue with a thin white stripe in the center and narrow white stripes at the edges. There were ten separate operations during this conflict; service in each of which entitled the recipient to one three-sixteenth inch bronze star or a three-sixteenth inch silver star in lieu of five bronze stars. Naval personnel who were attached to units of the Fleet Marine Force (FMF) during this campaign are entitled to wear a small Marine Corps emblem device on the ribbon.

The ten Navy campaign designations for the Korean Service Medal are:

North Korean aggression,
27 June 1950 - 2 November 1950

Communist China aggression,
3 November 1950 - 24 January 1951

Inchon Landing,
13 September - 17 September 1950

1st United Nations counteroffensive,
25 January 1951 - 21 April 1951

Communist China spring offensive,
22 April 1951 - 8 July 1951

United Nations summer-fall offensive,
9 July 1951 - 27 November 1951

2nd Korean winter,
28 November 1951 - 30 April 1952

Korean defensive, summer-fall 1952,
1 May 1952 - 30 November 1952

3rd Korean winter,
1 December 1952 - 30 April 1953

Korean summer 1953,
1 May 1953 - 27 July 1953

ANTARCTIC SERVICE MEDAL

Bronze, Gold, or Silver

Instituted: 1960
Dates: 1946 - Present
Criteria: 30 calendar days of service on the Antarctic Continent
Devices: Bronze, gold, silver disks (representing the bars below)
Bars: "Wintered Over" in bronze, gold, silver

For participation in an expedition, operation or support of a U.S. operation in Antarctic after 1 January 1946. The Antarctic Service Medal is worn after the Korean Service Medal and before the Armed Forces Expeditionary Medal.

The Antarctic Service Medal was established by an Act of Congress on 7 July 1960. The ribbon was approved in 1961 and the medal in 1963. The medal is awarded to any American or resident alien who subsequent to 1 January 1946 served on the Antarctic continent on or in support of U.S. operations there. Originally no minimum time was required for the medal; since 1 June 1973 a minimum of 30 days at sea or ashore south of sixty degrees latitude is required. One day on the continent counts for two days toward the thirty day eligibility.

The Antarctic Service Medal was designed by the U.S. Mint. The medal is a green-gold disc showing a man in the center with Antarctic clothing. On either side of the man is the raised inscription ANTARCTICA SERVICE. In the distant background is a mountain range and a line of clouds. The medal's reverse has the words COURAGE, SACRIFICE and DEVOTION set in the lines over a polar projection of the Antarctic continent. Around the edges is a border of penguins and marine life. The ribbon has a narrow white center stripe flanked on either side by progressively darker shades of blue and borders of black. Individuals who spend the months of March to October are entitled to wear a bronze clasp with the words WINTERED OVER on the suspension ribbon for the first stay, a gold clasp for the second stay and a silver clasp representing a third winter on the Antarctic continent. Five-sixteenth inch discs of the same finish are worn on the ribbon bar to represent the clasps.

UNITED STATES ANTARCTIC EXPEDITION MEDAL

Instituted: 1945
Dates: 1939 - 1941
Criteria: Participation in Admiral Byrd's Antarctic Expedition
Devices: None

For participation of the First United States Antarctic Expeditionary of 1939-1941.

The United States Antarctic Expedition Medal was authorized by Congress on 24 September 1945. The medal was awarded to officers and men who accompanied Admiral Byrd in the First United States Antarctic Expedition of 1939-1941. Although the Expedition was led by Admiral Byrd it was not named for him, since it was an official U.S. Government Operation.

The medal is a bronze disc with a map of the South Pole showing Antarctica, Little America and Palmerland. A scroll above the map has the words SCIENCE PIONERING EXPLORATION. The disc is circled by the words THE UNITED STATES - ANTARCTIC - EXPEDITION 1939 - 1941. The medal's reverse had a space for the recipients name to be engraved. The medal is suspended by a light blue ribbon with a wide white stripe bordered with thin red and white stripes.

ARMED FORCES EXPEDITIONARY MEDAL

Silver

Bronze

Instituted: 1961
Dates: 1958 to Present
Criteria: Participation in military operations not covered by a specific campaign medal
Devices: Bronze, silver star, bronze Marine Corps device
Notes: Authorized for service in Vietnam until establishment of Vietnam Service Medal

For participating in designated operations after I July 1958. The Armed Forces Expeditionary Medal is worn after the Antarctic Service Medal and before the Vietnam Service Medal.

The Armed Forces Expeditionary Medal was established by Executive Order on 4 December 1961. The medal is awarded to any member of the Armed Forces who participates in or is in support of U. S. Military operations, U.S. Military operations in support of the United Nations, or U.S. Military operations of assistance to friendly nations for which a specific campaign medal has not been established. A minimum of thirty consecutive days, or sixty nonconsecutive days are required for eligibility, unless the period of the operation was less than thirty days and in that case full participation during the operation is required. Additionally, personnel engaged in combat, or an equally hazardous duty, qualify for the award without regard to time in the area. The medal was initially awarded for Vietnam service between 1 July 1958 and 3 July 1965, prior to the establishment of the Vietnam Service Medal. An individual awarded this medal for this period of Vietnam service may either retain the medal or request the Vietnam Service Medal, but may not have both for this service.

The Armed Forces Expeditionary Medal was designed by the Army's Institute of Heraldry. The medal is a bronze disc with an American eagle, with wings raised, perched on a sword in front of a compass rose. The design is encircled by the words ARMED FORCES at the top and EXPEDITIONARY SERVICE at the bottom. These words are separated by sprigs of laurel. The reverse of the medal has the shield which appears on the Presidential seal, encircled with branches of laurel at the bottom and the raised inscription UNITED STATES OF AMERICA at the top. The ribbon is light blue with three narrow stripes of red, white and blue in the center and borders of black, brown, yellow and green. Three-sixteenth inch bronze stars are authorized for the current twenty-four qualifying operations. A three-sixteenth inch silver star is worn in lieu of five bronze stars. Naval personnel who were attached to units of the Fleet Marine Force (FMF) during campaigns awarded this medal are entitled to wear a small Marine Corps emblem device on the ribbon.

To date the Armed Forces Expeditionary Medal has been awarded for the following operations:

Lebanon, 1958
Taiwan Straights, 1958-1959
Quemoy & Matsu Islands, 1958-1963
Vietnam, 1958-1965
Congo, 1960-1962
Laos, 1961-1962
Berlin, 1961-1963
Cuba, 1962-1963
Congo, 1964
Dominican Republic, 1965-1966
Korea, 1966-1974
Cambodia, Thailand, 1973
Cambodia Evacuation, 1975
Mayaquez Operation, 1975
Vietnam Evacuation, 1975
El Salvador, 1981-1992
Grenada, 1983
Lebanon, 1983-1987
Libia, 1986
Persian Gulf, 1987-1990
Panama, 1989-1990
Somalia, 1992-1995
Haiti, 1994-1995
Joint Endeavor (former Yugoslavia), 1995-1996
Joint Guard (former Yugoslavia), 1996-1998
Joint Forge (former Yugoslavia), 1998-TBD

Iraqi Operations:
Former Republic of Yugoslavia -
Specific Operations, 1995-TBD
Iraq Operations -
Specific Operations, 1995-TBD

VIETNAM SERVICE MEDAL

Instituted: 1965
Dates: 1965-73
Criteria: Service in Vietnam, Laos, Cambodia or Thailand during the above period
Devices: Bronze, silver star, bronze Marine Corps device

For service in Southeast Asia and contiguous waters or airspace during the Vietnam War. The Vietnam Service Medal is worn after the Armed Forces Expeditionary Medal and before the Southwest Asia Service Medal.

The Vietnam Service Medal was established by Executive Order on 8 July 1965. The medal was awarded to all members of the Armed Forces who served in Vietnam and contiguous waters and airspace from 3 July 1965 to 28 March 1973. In addition, individuals serving in Laos, Thailand, or Cambodia in direct support of operations in Vietnam during the same period are also eligible. Individuals previously awarded the Armed Forces Expeditionary Medal for service in Vietnam between July 1958 and July 1965 may, upon request, exchange that medal for the Vietnam Service Medal.

The Vietnam Service Medal was designed by Thomas Jones, formerly a sculptor with the Army's Institute of Heraldry. The medal is a bronze disc with a dragon behind a grove of bamboo trees. Below the design is the three line raised inscription REPUBLIC OF VIETNAM SERVICE. The reverse of the medal shows a crossbow below a torch with the raised inscription UNITED STATES OF AMERICA encircling the lower edge. The ribbon is yellow with three narrow red stripes in the center and narrow green stripes at the borders. Three-sixteenth inch bronze stars denote participation in each one of the seventeen Vietnam campaigns with a three-sixteenth inch silver star worn in lieu of five bronze stars. Naval personnel who were attached to units of the Fleet Marine Force (FMF) during this campaign are entitled to wear a small Marine Corps emblem device on the ribbon.

The seventeen Navy campaign designations for the Vietnam Service Medal are:

Vietnam Advisory Campaign,
15 March 1962 - 7 March 1965

Vietnam Defense Campaign,
8 March 1965 - 24 December 1965

Vietnam Counteroffensive,
25 December 1965 - 30 June 1966

Vietnam Counteroffensive Phase II,
1 July 1966 - 31 May 1967

Vietnam Counteroffensive Phase III,
1 June 1967 - 29 January 1968

TET Counteroffensive,
30 January 1968 - 1 April 1968

Vietnam Counteroffensive Phase IV,
2 April 1968 - 30 June 1968

Vietnam Counteroffensive Phase V,
1 July 1968 - 1 November 1968

Vietnam Counteroffensive Phase VI,
2 November 1968 - 22 February 1969

TET '69 Counteroffensive,
23 February 1969 - 8 June 1969

Vietnam Summer -Fall 1969,
9 June 1969 - 31 October 1969

Vietnam Winter - Spring 1970,
1 November 1969 - 30 April 1970

Sanctuary Counteroffensive,
1 May 1970 - 30 June 1970

Vietnam Counteroffensive Phase VII,
1 July 1970 - 30 June 1971

Consolidation I,
1 July 1971 - 30 November 1971

Consolidation II,
1 December 1971 - 29 March 1972

Vietnam Cease-fire Campaign,
30 March 1972 - 28 January 1973

SOUTHWEST ASIA SERVICE MEDAL

Instituted: 1991 Dates: 1990-1995
Criteria: Active participation in, or support of, Operations Desert Shield and/or Desert Storm
Devices: Bronze star, bronze Marine Corps device
Notes: Terminal date of service was 30 November 1995

For service in Southwest Asia during Desert Shield/Desert Storm operations, or what is often referred to as the Persian Gulf War. The Southwest Asia Service Medal is worn after the Vietnam Service Medal and before the Kosovo Service Medal.

The Southwest Asia Service Medal was established by Executive Order on 12 March, 1991. The medal is awarded to all members of the Armed Forces who participated in military operations, or in direct support of military operations in Southwest Asia and contiguous waters and airspace from 2 August 1990 to 30 November 1995.

The Southwest Asia Service Medal was designed by the Army's Institute of Heraldry. The medal is a circular bronze disc with the upper part of the medal showing a desert setting with a rising sun, tent, troops, armored personnel carrier and helicopter. The lower portion of the medal shows a sea setting with clouds, a warship and two fixed wing aircraft. The two settings are separated by the raised inscription SOUTHWEST ASIA SERVICE set in two lines. The reverse of the medal has a sword entwined with a palm leaf pointing up encircled by the raised inscription UNITED STATES OF AMERICA. The ribbon is tan with black borders and a thin black stripe in the center flanked on either side by stripes of green. Between the center stripes and the borders are narrow stripes of red, white and blue. One three-sixteenth bronze star is worn for participation in each of the three authorized campaigns. Naval personnel who were attached to Marine units of the Fleet Marine Force (FMF) during this campaign are entitled to wear a small Marine Corps emblem device on the ribbon.

The three Navy campaign designations for the Southwest Asia Service Medal are:

Defense of Saudi Arabia,	Liberation and Defense of Kuwait,	Cease-fire,
2 August 1990 - 16 January 1991	17 January 1991 - 11 April 1991	12 April 1991 - 30 November 1995

KOSOVO CAMPAIGN MEDAL

Bronze

Instituted: 2000
Dates: 1999 - To be determined
Criteria: Active participation in, or in direct support of Kosovo operations
Devices: Bronze star, bronze Marine Corps device

For participation in, or in direct support of Kosovo operations. The Kosovo Campaign Medal is worn after the Southwest Asia Service Medal and before the Armed Forces Service Medal.

The Kosovo Campaign Medal was established by Executive Order on 15 May 2000. The medal is awarded to all members of the Armed Forces who participated in, or provided direct support to, Kosovo operations within established areas of eligibility (AOE) from 24 March 1999 to a date yet to be determined. The service member must have been a member of a unit participating in or engaged in support of the operation for 30 consecutive days or 60 nonconsecutive days.

The Kosovo Campaign Medal was designed by the Army's Institute of Heraldry. The medal is a circular bronze disc with a depiction of a rising sun behind a Dinartic Alpine mountain pass and a fertile valley over a stylized wreath of grain with the inscription KOSOVO CAMPAIGN. The reverse of the medal has an outline of the Province of Kosovo over the NATO star and the inscription IN DEFENSE OF HUMANITY. The ribbon is blue and red with three narrow red, white and blue stripes in the center. One three-sixteenth bronze star is worn for participation in each of the authorized campaigns. Naval personnel who were attached to Marine units of the Fleet Marine Force (FMF) during this campaign are entitled to wear a small Marine Corps emblem device on the ribbon.
The Navy campaign designations for the Kosovo Campaign Medal as of this date are:

Kosovo Air Campaign	Kosovo Defense Campaign
24 March 1999 - 10 June 1999	10 June 1999 - TBD

ARMED FORCES SERVICE MEDAL

Silver

Bronze

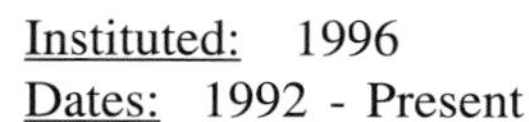

Instituted: 1996
Dates: 1992 - Present
Criteria: Participation in military operations not covered by a specific war medal or the Armed Forces Expeditionary Medal
Devices: Bronze, silver star

For participating in peaceful military operations deemed significant on or after 1 June 1992. The Armed Forces Service Medal is worn after the Kosovo Service Medal and before the Humanitarian Service Medal.

The Armed Forces Service Medal was authorized by Executive Order on 11 January 1996. The medal is awarded to any member of the Armed Forces who participates in a military operation deemed to be a significant activity in which no foreign armed opposition or hostile action is encountered and for which no other U.S. service medal is authorized. Qualification includes at least one day of participation in the designated area. Direct support of the operation and aircraft flights within the area also qualify for this award as long as at least one day is served within the designated area.

The Armed Forces Service Medal was designed by the Army's Institute of Heraldry. The medal is a bronze disc which displays the torch from the Statue of Liberty in the center, encircled by the raised inscription ARMED FORCES SERVICE MEDAL. The reverse of the medal shows the eagle from the Seal of the Department of Defense, encircled by a laurel wreath and the raised inscription IN PURSUIT OF DEMOCRACY. The ribbon has a wide light blue center stripe flanked by narrow stripes of tan, olive green and dark green. The borders are narrow stripes of tan. Three-sixteenth inch bronze stars denote successive awards of this medal.

HUMANITARIAN SERVICE MEDAL

Silver

Bronze

Instituted: 1977
Dates: 1975 - Present
Criteria: Direct participation in specific operations of a humanitarian nature
Devices: Bronze, silver star

For direct participation in specific operations of a humanitarian nature subsequent to 1 April 1975. The Humanitarian Service Medal is worn after the Armed Forces Service Medal and before the Military Outstanding Volunteer Service Medal.

The Humanitarian Service Medal was authorized by Executive Order on 19 January 1977. The medal was established to honor members of the Armed Forces who distinguish themselves by meritorious direct participation in a significant military operation of a humanitarian nature, or rendered a service to mankind. The participation must be "hands on" at the site of the operation. Recent qualifying operations include the Oklahoma City Bombing Disaster Relief Operation (19 April - 3 May 1995) and Southeast Flood Disaster Relief Operation (July - August 1994).

The Humanitarian Service Medal was designed by the Army's Institute of Heraldry. The medal is a bronze disc which displays a right hand open and pointing upward within a circle. At the top of the medal's reverse is the raised inscription FOR HUMANITARIAN SERVICE set in three lines. In the center is an oak branch with three acorns and leaves and below this, around the edge, is the raised inscription UNITED STATES ARMED FORCES. The ribbon is medium blue with a wide center stripe of navy blue. The ribbon is edged by a wide stripe of purple with a narrow stripe of white inboard. Additional awards are denoted by three-sixteenth inch bronze stars.

MILITARY OUTSTANDING VOLUNTEER SERVICE MEDAL

Silver Bronze

Instituted: 1993 Dates: 1993 - Present
Criteria: Awarded for outstanding and sustained voluntary service to the civilian community
Devices: Bronze, silver star

For outstanding and sustained voluntary service to a civilian community. The Military Outstanding Volunteer Service Medal is worn after the Humanitarian Service Medal and before the Armed Forces Reserve Medal when medals are worn and before the Navy Sea Service Deployment Ribbon on a ribbon display.

The Military Outstanding Volunteer Service Medal was established by Executive Order on 9 January 1993. The medal is intended to recognize members of the Armed Forces who perform outstanding volunteer service to a civilian community. The service performed must be strictly voluntary and not duty-related and must reflect sustained direct individual involvement in the volunteer activity (the medal is not intended to recognize a single act). Both Active Duty and Reserve personnel are eligible.

The Military Outstanding Volunteer Service Medal was designed by the Army's Institute of Heraldry. The medal is a bronze disc with a five pointed star in the center. The star has rings around each point and is encircled with a stylized laurel wreath. The reverse of the medal has an oak leaf branch with three oak leaves and two acorns and the raised inscription OUTSTANDING VOLUNTEER SERVICE set in three lines. Around the bottom edge are the words UNITED STATES OF AMERICA. The ribbon is medium blue with two yellow stripes in the center, each bordered with thin green stripes. Additional yellow stripes (not bordered) are near the edges. Additional awards are denoted by three-sixteenth inch bronze stars.

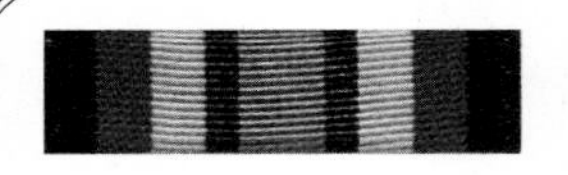

NAVY SEA SERVICE DEPLOYMENT RIBBON

Silver

Bronze

Instituted: 1981 Dates: Retroactive to 1974
Criteria: 12 months accumulated active duty on deployed vessels, operating away from their home port for extended periods, or at an overseas duty station
Devices: Bronze, silver star

To recognize the unique and demanding nature of sea service and the arduous duty attendant with deployment. The Navy Sea Service Deployment Ribbon is worn after the Military Outstanding Volunteer Service Medal and before the Navy Arctic Service Ribbon.

The Navy Sea Service Deployment Ribbon was approved by the Secretary of the Navy in 1981 and made retroactive to 15 August 1974. The ribbon was created to recognize the unique and demanding nature of sea service and the arduous duty attendant with such service deployments and overseas service. The award is made for twelve months of accumulated sea duty or 12 months accumulated duty at an overseas sea duty station.

The Navy Sea Service Deployment Ribbon is a ribbon only award. The ribbon consists of a wide center stripe of light blue, bordered on either side by a narrow stripe of medium blue and equal stripes of gold, red and navy blue. Additional awards are denoted by three-sixteenth inch bronze stars.

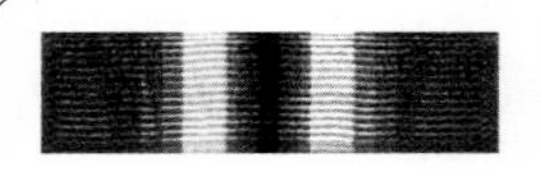

NAVY ARCTIC SERVICE RIBBON

Instituted: 1986
Criteria: 28 days of service on Naval vessels operating above the Arctic Circle
Devices: None

For participation in operations in support of the Arctic Warfare Program. The Navy Arctic Service Ribbon is worn after the Navy Sea Service Deployment Ribbon and before the Naval Reserve Sea Service Ribbon.

The Navy Arctic Service Ribbon was authorized by the Secretary of the Navy on 8 May 1986 and established by an OPNAVNOTE (Chief of Naval Operations Notice) on 3 June 1987. The ribbon is awarded to members of the Naval Service who participate in operations in support of the Arctic Warfare Program. To be eligible the individual must have served 28 days north of, or within 50 miles of the Marginal Ice Zone (MIZ). The MIZ is defined as an area consisting of more than 10% ice concentration.

The Navy Arctic Service Ribbon is a ribbon only award. The ribbon is medium blue with a narrow center stripe of navy blue flanked on either side by three thin stripes of gradually lighter shades of blue, a narrow stripe of white, followed again by two thin stripes of gradually darker shades of blue. There are no provisions for additional awards.

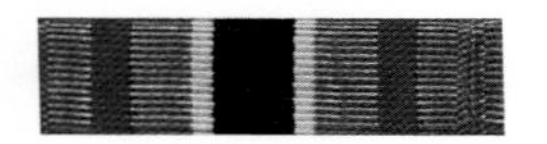

NAVAL RESERVE SEA SERVICE RIBBON

Instituted: 1987
Criteria: 24 months of cumulative service embarked on Naval Reserve vessels or embarked Naval Reserve unit
Devices: Bronze, silver star

For Naval Reservists who have 24 months of cumulative service embarked on Naval Reserve vessels or part of an embarked Naval Reserve unit. The Naval Reserve Sea Service Ribbon is worn after the Navy Arctic Service Ribbon and before the Navy and Marine Corps Overseas Service Ribbon.

The Naval Reserve Sea Service Ribbon was instituted in 1987. The Naval Reserve Sea Service Ribbon is a light blue ribbon with a wide center stripe of navy blue bordered with thin gold stripes and bands of red near the ribbon borders. Additional awards are denoted by three-sixteenth bronze stars.

NAVY AND MARINE CORPS OVERSEAS SERVICE RIBBON

Instituted: 1987
Criteria: 12 months consecutive or accumulated duty at an overseas shore base duty station
Dates: 1974 to present
Devices: Bronze, silver star

For 12 months consecutive or accumulated active duty at an overseas duty station; or 30 consecutive days or 45 cumulative days of active duty for training or temporary active duty. The Navy and Marine Corps Overseas Service Ribbon is worn after the Navy Arctic Service Ribbon and before the Navy Recruiting Service Ribbon.

The Navy and Marine Corps Overseas Service Ribbon was approved by the Secretary of the Navy and authorized by an OPNAVNOTE (Chief of Naval Operations Notice) on 3 June 1987 and made retroactive to 15 August 1974. The award is made to Active Duty Members of the Naval Service who serve 12 months at an overseas duty station.

The Navy and Marine Corps Overseas Service Ribbon is a ribbon only award. The ribbon has a wide red center stripe, bordered on either side by a thin yellow stripe, a wide navy blue stripe, a thin yellow stripe and a narrow medium blue border. Additional awards are denoted by three-sixteenth inch bronze stars.

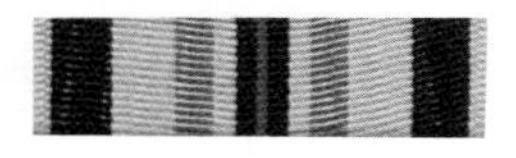

NAVY RECRUITING SERVICE RIBBON

Instituted: 1989
Dates: 1974- Present
Criteria: 3 successful consecutive years of recruiting duty
Devices: Bronze star

For successful completion of three consecutive years of recruiting duty. The Navy Recruiting Service Ribbon is worn after the Navy and Marine Corps Overseas Service Ribbon and before the Navy Recruit Training Service Ribbon.

The Navy Recruiting Service Ribbon was established in 1989 and made retroactive to 15 August 1974.

The Navy Recruiting Service Ribbon is a ribbon only award. The ribbon is gold with Navy blue stripes near the borders and at the center. The blue center stripe has a thin red stripe in the center and is bordered by stripes of light green on either side. Additional awards are denoted by three-sixteenth inch bronze stars.

NAVY RECRUIT TRAINING SERVICE RIBBON

Bronze

Instituted: 1998 Dates: 1995 - Present
Criteria: Successful enlisted tour of duty as a RDC assigned to RTC
Devices: Bronze star

For enlisted personnel who have served successfully as Recruit Division Commanders (RDC) assigned to Recruit Training Command (RTC), Naval Training Center, Great Lakes. The Navy Recruit Training Service Ribbon is worn after the Navy Recruiting Service Ribbon and before the Armed Forces Reserve Medal.

The Navy Recruit Training Service Ribbon was established in 1998 and made retroactive to tours completed after 1 October 1995. The tour of duty as RDC must have completed at least five Divisions trained over a minimum tour of three years.

The Navy Recruit Training Service Ribbon is a ribbon only award. The ribbon is Navy blue with a wide red center stripe with borders of gold. Additional awards are denoted by three-sixteenth inch bronze stars.

ARMED FORCES RESERVE MEDAL

Bronze, Silver, Gold Hourglass

Bronze

Instituted: 1950 Dates: 1949 - Present
Criteria: 10 years of honorable service in any reserve component of the United States Armed Forces
Devices: Bronze, silver and gold hourglass, bronze numeral and letter "M"

For ten years of honorable service in Reserve components of the Armed Forces. The Armed Forces Reserve Medal is worn after the Navy Recruit Training Service Ribbon and before foreign medals when medals are worn and after the Navy Recruiting Service Ribbon

The Armed Forces reserve Medal was authorized by Executive Order on 25 September 1950 and amended on 19 March 1952. The medal is awarded to individuals who complete ten years of honorable satisfactory service in any of the reserve components, including the National Guard. The service need not be continuous, but the ten years must be completed within a twelve year period. Satisfactory service is defined as having been credited with a minimum of fifty reserve retirement points.

The Armed Forces Reserve Medal was designed by the Army's Institute of Heraldry. The medal is a bronze disc depicting a vertical flaming torch centered over a bugle and powder horn. In the background are thirteen stars and rays representing the thirteen original colonies. The reverse has the Navy emblem around which is the raised circular inscription ARMED FORCES RESERVE. The ribbon is buff with a narrow medium blue center stripe and three thin medium blue stripes on each edge.

In certain cases, an early award of the Armed Forces Reserve Medal may be made. If a reservist is called to active duty prior to the completion of the prescribed ten years of Reserve service, the medal is awarded immediately with a bronze letter "M" to denote the mobilization. Upon completion of the ten year period, reservists wear the Armed Forces Reserve Medal with a bronze hourglass device. Silver and gold hourglass devices are awarded at the end of twenty and thirty years of reserve service, respectively. Since only one letter "M" device may be awarded, a bronze numeral denotes the number of times the recipient has been mobilized for active duty.

NAVAL RESERVE MEDAL (OBSOLETE)

Instituted: 1938 Dates: 1938 - Present
Criteria: 10 years of honorable service in the U.S. Naval Reserve
Devices: None

For Naval Reservists who fulfilled ten years of honorable service in the U.S. Naval Reserve prior to 1958.

The Naval Reserve Medal was established 12 September 1938 and replaced by the Armed Forces Reserve Medal in 1950. Personnel who were eligible for the medal from 1950 to 1958 could choose which award to receive.

The medallion of the Naval Reserve Medal as a bronze disc showing an eagle perched on an anchor with a sea and the rays of the sun in the background. The reverse of the medal has the words UNITED STATES NAVAL RESERVE encircling the outer edge. In the center are the words: FAITHFUL SERVICE, centered on two lines and at the bottom is a five pointed star. The ribbon is red with thin border stripes of gold and blue.

FOREIGN AWARDS

This is where a foreign decoration or awards go in order of precedence.

Foreign awards are worn in the order of their receipt, unless there are two or more awards from the same country, at which time they are worn in accordance with the country's precedence.

In order to retain and wear a specific foreign awards, a letter must be sent to CNO N09B33 for approval.

The example shown is the Republic of Vietnam Armed Forces Honor Medal.

PHILIPPINE REPUBLIC PRESIDENTIAL UNIT CITATION

Instituted: 1948
Criteria: Awarded to units of the U.S. Armed Forces for service in the war against Japan and/or for 1970 and 1972 disaster relief
Devices: Bronze star

For service in a unit cited in the name of the President of the Republic of the Philippines for outstanding performance in action.

The Philippine Republic Presidential Unit Citation is a foreign award. It was awarded to members of the Armed Forces of the United States for services during the defense of the liberation of the Philippines during World War II. The award was made in the name of the President of the Republic of the Philippines. The Philippine Presidential Unit Citation was also awarded to U.S. Forces who participated in disaster relief operations 1 August 1970 to 14 December 1970 and 21 July 1972 to 15 August 1972.

The Philippine Republic Presidential Unit Citation is a ribbon only award. The ribbon has three wide stripes of blue, white and red enclosed in a rectangular one-sixteenth inch gold frame with laurel leaves. A three sixteenth inch bronze star is authorized for a recipient who received more than award.

REPUBLIC OF KOREA PRESIDENTIAL UNIT CITATION

Instituted: 1951
Criteria: Awarded to certain units of the U.S. Armed Forces for services rendered during the Korean War
Devices: None

Awarded by the Republic of Korea for service in a unit cited in the name of the President of the Republic of Korea for outstanding performance in action.

The Republic of Korea Presidential Unit Citation is a Foreign Unit Award. It was awarded to units of the United Nations Command for service in Korea during the Korean War from 1950 to 1954. The award was made in the name of the President of the Republic of Korea.

The Republic Korea Presidential Unit Citation is a ribbon only award. The ribbon is white bordered with a wide green stripe and thin stripes of white, red, white, red, white and green. In the center is an ancient oriental symbol called a tae-guk (the top half is red and the bottom half is blue). The ribbon is enclosed in a rectangular one-sixteenth inch gold frame with laurel leaves.

REPUBLIC OF VIETNAM PRESIDENTIAL UNIT CITATION (FRIENDSHIP RIBBON)

Instituted: 1954
Criteria: Awarded to certain units of the U.S. Armed Forces for humanitarian service in the evacuation of civilians from North and Central Vietnam
Devices: None

Awarded by the Republic of Vietnam for service in a unit cited in the name of the President of the Republic of Vietnam for outstanding performance in action.

The Republic of Vietnam Presidential Unit Citation is a Foreign Unit Award. Referred to as the “Friendship Ribbon”, it was awarded to members of the United States Military Assistance Advisory Group in Indo-China for services rendered during August and September 1954. The ribbon is awarded in the name of the President of the Republic of Vietnam.

The Republic of Vietnam Presidential Unit Citation is a ribbon only award. The ribbon is yellow with three narrow red stripes in the center. The ribbon is enclosed in a rectangular one-sixteenth inch gold frame with laurel leaves.

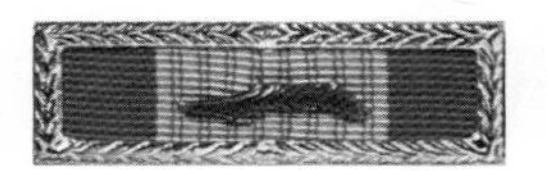

REPUBLIC OF VIETNAM GALLANTRY CROSS UNIT CITATION

Bronze Palm

Instituted: 1966
Criteria: Awarded to certain units of the U.S. Armed Forces for valorous combat achievement during the Vietnam War, 1 March 1961 to 28 March 1974
Devices: Bronze palm

Awarded by the Republic of Vietnam to units of the U.S. Armed Forces in recognition of valorous achievement in combat during the Vietnam War (1 March 1961-28 March 1974).

The Republic of Vietnam Gallantry Cross Unit Citation was established on 15 August 1950. The ribbon was awarded to units by the Republic of Vietnam for the same services as would be required for the Navy Unit Commendation.

The Republic of Vietnam Gallantry Cross Unit Citation is a foreign award. The ribbon is red with a very wide yellow center stripe, which has eight very thin double red stripes. The ribbon bar is enclosed in a one-sixteenth inch gold ssssframe with laurel leaves. A bronze palm device is attached to the ribbon.

REPUBLIC OF VIETNAM CIVIL ACTIONS UNIT CITATION

Bronze Palm

Instituted: 1966
Criteria: Awarded to certain units of the U.S. Armed Forces for meritorious service during the Vietnam War, 1 March 1961 to 28 March 1974
Devices: Bronze palm

Awarded by the Republic of Vietnam to units in recognition of meritorious civil action service.

The Republic of Vietnam Civil Actions Unit Citation, sometimes called the Civic Actions Honor Medal, was widely bestowed on American forces in Vietnam as a unit citation for which there is no medal authorized. The award recognizes outstanding achievements made by units in the field of civil affairs.

The Republic of Vietnam Civil Actions Unit Citation is a foreign award. The ribbon is enclosed in a rectangular one-sixteenth inch gold frame with laurel leaves and is normally awarded with a bronze laurel leaf palm attachment.

PHILIPPINE DEFENSE MEDAL AND RIBBON

Bronze

Institute: 1945

Criteria: Service in defense of the Philippines between 8 December 1941 and 15 June 1942

Devices: Bronze star

Awarded by the Philippine Commonwealth for service in defense of the Philippines.

The Philippine Defense Ribbon was authorized in 1944 by the United States and the Philippine Commonwealth government. The ribbon was awarded to members of the United States Armed Forces for service in the defense of the Philippines from 8 December 1941 to 15 June 1942.

The Philippine Defense Ribbon is a non-U.S. Service Award. Although not authorized for wear on the U.S. military uniform, a medal was designed and struck by the Philippine Government. The medal was designed by the Manila firm of El Oro. The medal is a gold disc with an outer edge of ten scallops. At the medal's center is a female figure with a sword and shield representing the Philippines.. Above the figure are three stars and surrounding it is a green enamel wreath. At the bottom right of the medal is a map of Corregidor and Bataan. At the bottom left is a floral design. The reverse of the medal has the raised inscription FOR THE DEFENSE OF THE PHILIPPINES set in four lines. The ribbon is red, with two white stripes near the edges and three white five-pointed stars in the center. A three-sixteenth inch bronze star denotes additional service during the prescribed eligibility period.

PHILIPPINE LIBERATION MEDAL AND RIBBON

Bronze

Instituted: 1944

Criteria: Service in the liberation of the Philippines between 17 October 1944 and 3 September 1945

Devices: Bronze star

Awarded by the Philippine Commonwealth for the liberation of the Philippines.

The Philippine Liberation Ribbon was authorized for members of the Naval Service by ALNAV 64 on 5 April 1945. The ribbon was awarded to members of the United States Armed Forces for service in the liberation of the Philippines from 17 October1944 to 3 September 1945.

The Philippine Liberation Medal is a non-U.S. Service Award. Although not authorized for wear on the U.S. military uniform, a medal was designed and struck by the Philippine Government The medal was designed by the Manila firm of El Oro. The medal is gold with a Philippine sword, point up, superimposed over a white native shield. The shield has three gold stars at the top and the word LIBERTY below. Below are vertical stripes of blue, white and red enamel with the sword being in the center of the white stripe. At the sides of the medal and below the shield are gold arched wings. The reverse of the medal has the raised inscription FOR THE LIBERATION OF THE PHILIPPINES set in four lines. The ribbon is red with narrow blue stripe and a narrow white stripe in the center. Three-sixteenth inch bronze stars (up to two) denote additional service during the prescribed eligibility period.

PHILIPPINE INDEPENDENCE MEDAL

Instituted: 1946 (Army: 1948)
Criteria: Receipt of both the Philippine Defense and Liberation Medals/Ribbons. Originally presented to those present for duty in the Philippines on 4 July 1946
Devices: None

Awarded by the Philippine Commonwealth to those members of the Armed Forces who received both the Philippine Defense Ribbon and the Philippine Liberation Ribbon.

The Philippine Independence Ribbon was authorized in 1946 by the United States and the government of the Philippine Commonwealth. The ribbon was presented to those members of the United States Armed Forces who were serving in the Philippines on 4 July 1946 or who were previously awarded the Philippine Defense Ribbon and the Philippine Liberation Ribbon.

The Philippine Independence Ribbon is a Non-U.S. Service Award. Although not authorized to be worn on the U.S. military uniform, a medal was designed and struck by the Philippine Government. The medal was designed by the Manila firm of El Oro. The medal is a gold disc with a female figure in the center, dressed in native garb and holding the Philippine flag. There are flags on either side of the figure and she is surrounded by a circular border. Inside the border is a raised inscription PHILIPPINE INDEPENDENCE around the top and July 4 1946 at the bottom. The ribbon is medium blue with a narrow white center stripe bordered by thin red stripes. There are thin yellow stripes at each edge.

UNITED NATIONS (KOREAN) SERVICE MEDAL

Instituted: 1951
Criteria: Service on behalf of the United Nations in Korea between 27 June 1950 and 27 July 1954
Devices: None
Notes: Above date denotes when award was authorized for wear by U.S. military personnel

For service on behalf of the United Nations in Korea during the Korean War.

The United Nations Service Medal is a Non-U.S. Service Award. It was authorized by the United Nations General Assembly on 12 December 1950 and the Department of Defense approved it for wear by members of the United States Armed Forces on 27 November 1951. The medal was awarded to any member of the United States Armed Forces for service in support of the United Nations Command during the period from 27 June 1950 to 27 July 1954. Individuals who were awarded the Korean Service Medal automatically established eligibility for this decoration.

The United Nations Service Medal was designed by the United Nations. The medal is a bronze disc with the United Nations emblem (a polar projection of the world taken from the North Pole, encircled by two olive branches). The reverse of the medal has the raised inscription FOR SERVICE IN DEFENSE OF THE PRINCIPLES OF THE CHARTER OF THE UNITED NATIONS set in five lines. The medal is suspended permanently from a bar, similar to British medals, with the raised inscription KOREA. The ribbon passes through the bar and consists of 17 narrow stripes of alternating light blue and white.

UNITED NATIONS (OBSERVER) MEDAL

Bronze

Institituted: 1964
Criteria: 6 months service with any authorized U.N. operation:
Devices: Bronze star
Notes: Above date denotes when award was authorized for wear by U.S. military personnel

For six months service on behalf of the United Nations in one of eleven operations. The United Nations Service Medal is a non-U.S. Service Award. It was authorized by the United Nations General Assembly on 30 July 1959 and approved by Executive Order on 11 March 1964. The medal was awarded to any member of the United States Armed Forces for not less than six months service in support of a United Nations operation. There are a total of eleven different operations involving members of the Armed Forces of the United States as of this writing.

The United Nations Service Medal was designed by the United Nations. The medal is a bronze disc with the United Nations emblem (a polar projection of the world taken from the North Pole, encircled by two olive branches). Centered above this are the letters UN. The reverse of the medal has the raised inscription IN THE SERVICE OF PEACE. The medallion for all UN operations is the same for all authorized operations. The basic ribbon is United Nations blue with narrow stripes of white near the edges; however, each authorized operation has a unique ribbon. Individuals who have participated in more than one UN operation wear only the first medal for which they qualify with a three-sixteenth inch bronze star for each subsequent award.

UNITED NATIONS MEDALS Summary shown on next two pages (110-111)

NATO MEDAL

Bronze

Instituted: 1992
Criteria: 30 days service in or 90 days outside the former Republic of Yugoslavia and the Adriatic Sea under NATO command in direct support of NATO operations
Devices: None
Notes: Above date denotes when award was authorized for wear by U.S. military personnel.
Former "Yugoslavia" Bar is not authorized for wear by U.S. military personnel. 

Awarded by The North Atlantic Treaty Organization (NATO) to military personnel who have served under NATO command or in direct support of NATO operations.

The NATO Medal is considered the same as a non-U.S. Service Award. The medal is awarded to members of the U.S. Armed Forces who served 30 days under NATO command, or 90 days in direct support of NATO operations in the Republic of Yugoslavia and the Adriatic Sea, or as designated by SACEUR from 1 July 1992 to a date currently undetermined.

The NATO Medal is a bronze disc with a raised edge. In the center of the medal is the NATO star surrounded by a wreath of olive leaves. The reverse of the medal contains the raised inscription IN SERVICE OF PEACE AND FREEDOM in both English (on top) and in French (below). Around the edge is a raised double ring with the enclosed inscription NORTH ATLANTIC TREATY ORGANIZATION in both English (on top) and in French (below). The ribbon is navy blue with narrow white stripe near the edges. The medal is sometimes issued with a clasp, which it is not worn by U.S. Armed Forces. Additional awards are denoted by three sixteenth-inch bronze stars.

United Nations Missions Participated in by United States Armed Forces Personnel

Until recently, U.S. military personnel serving on or with a United Nations mission were permitted to wear only two UN medals, the United Nations Korean Service Medal and the UNTSO (UN Truce Supervision Organization) Medal.

A change in Department of Defense regulations now authorizes military personnel to wear the ribbon of one of 11 UN missions. Only one UN ribbon may be worn. Subsequent mission awards are denoted by three-sixteenth inch bronze stars on the originally-earned UN ribbon. The United States has participated in 17 UN Missions (as of the date this book was published). To date, U.S. military personnel are permitted to wear medals from 11 of these missions as well as the United Nations Special Service Medal. The authorized medals are shown below:

1. KOREA - United Nations Korean Service
COUNTRY/LOCATION: Korea DATES: June 1950 to July 1953
COUNTRIES PARTICIPATING: (19) Australia, Belgium, Canada, Colombia, Ethiopia, France, Greece, Luxembourg, Netherlands, New Zealand, Philippines, South Korea, Thailand, Turkey, Union of South Africa, United Kingdom, United States (plus Denmark and Italy which provided medical support)
MAXIMUM STRENGTH (approx): 1,000,000 (UN & South Korea combined)
CURRENT STRENGTH: ------FATALITIES (approx): Korea: 415,000, U.S.: 55,000, Other UN: 3,100 CLASP(S): None

2. UNTSO - United Nations Truce Supervision Organization
COUNTRY/LOCATION: Palestine/Israel DATES: June 1948 to present
COUNTRIES PARTICIPATING: (20) Argentina, Australia, Austria, Belgium, Canada, Chile, China, Denmark, Finland, France, Ireland, Italy, Myanmar, Netherlands, New Zealand, Norway, Sweden, Switzerland, United States, USSR
MAXIMUM STRENGTH: 572 military observers (1948)
CURRENT STRENGTH: 143 (1999)
FATALITIES: 38 (1996)

3. UNMOGIP - United Nations Military Observer Group in India and Pakistan
COUNTRY/LOCATION: India, Pakistan (Jammu & Kashmir)
DATES: January 1949 to present
COUNTRIES PARTICIPATING: (15) Australia, Belgium, Canada, Chile, Denmark, Ecuador, Finland, Italy, Korean Republic, Mexico, New Zealand, Norway, Sweden, United States, Uruguay
MAXIMUM STRENGTH: 102 military observers (1965)
CURRENT STRENGTH: 45 (1999) FATALITIES: 9 (1996) CLASP(S): None

4. UNSF - United Nations Security Force in West New Guinea (West Irian) UNTEA-United Nations Temporary Executive Authority
COUNTRY/LOCATION: West Irian (West New Guinea)
DATES: October 1962 to April 1963
COUNTRIES PARTICIPATING: (9) Brazil, Canada, Ceylon, India, Ireland, Nigeria, Pakistan, Sweden, United States
MAXIMUM STRENGTH: 1,576 military observers (1963)
STRENGTH AT WITHDRAWAL: 1,576 FATALITIES: None CLASP(S): None

5. UNIKOM - United Nations Iraq-Kuwait Observation Mission
COUNTRY/LOCATION: Iraq, Kuwait DATES: April 1991 to present
COUNTRIES PARTICIPATING: (36) Argentina, Austria, Bangladesh, Canada, Chile, China, Denmark, Fiji, Finland, France, Germany, Ghana, Greece, Hungary, India, Indonesia, Ireland, Italy, Kenya, Malaysia, Nigeria, Norway, Pakistan, Poland, Romania, Russian Federation, Senegal, Singapore, Sweden, Switzerland, Thailand, Turkey, United Kingdom, United States, Uruguay, Venezuela
MAXIMUM STRENGTH: 1,187 military observers CURRENTSTRENGTH: 1,102 (1999) FATALITIES: 6 (1996)
CLASP(S): None

6.MINURSO - United Nations Mission for the Referendum in Western Sahara
COUNTRY/LOCATION: Western Sahara (Morocco) DATES: September 1991 to present
COUNTRIES PARTICIPATING: (36) Argentina, Australia, Austria, Bangladesh, Belgium, Canada, China, Egypt, El Salvador, Finland, France, Germany, Ghana, Greece, Guinea-Bissau, Honduras, Hungary, Ireland, Italy, Kenya, Korean Republic, Malaysia, Nigeria, Norway, Pakistan, Peru, Poland, Portugal, Russian Federation, Switzerland, Togo, Tunisia, United Kingdom, United States, Uruguay, Venezuela
MAXIMUM STRENGTH: 3,000 authorized (1,700 military observers and troops, 300 police officers, approx. 1,000 civilian personnel)
CURRENT STRENGTH: 316 (1999) FATALITIES: 7 (1996) CLASP(S): None

7. UNAMIC - United Nations Advance Mission in Cambodia **COUNTRY/LOCATION: Cambodia**
DATES: November 1991 to March 1992
COUNTRIES PARTICIPATING: (24) Algeria, Argentina, Australia, Austria, Belgium, Canada, China, France, Germany, Ghana, India, Indonesia, Ireland, Malaysia, New Zealand, Pakistan, Poland, Russian Federation, Senegal, Thailand, Tunisia, United Kingdom, United States, Uruguay
MAXIMUM STRENGTH: 1,090 military and civilian personnel (1992)
STRENGTH AT TRANSITION TO UNTAC: 1,090 FATALITIES: None

8. UNPROFOR - United Nations Protection Force
COUNTRY/LOCATION: Former Yugoslavia (Bosnia, Herzegovina, Croatia, Serbia, Montenegro, Macedonia)
DATES: March 1992 to December 1995
COUNTRIES PARTICIPATING: (43) Argentina, Australia, Bangladesh, Belgium, Brazil, Canada, Colombia, Czech Republic, Denmark, Egypt, Finland, France, Germany, Ghana, India, Indonesia, Ireland, Jordan, Kenya, Lithuania, Luxembourg, Malaysia, Nepal, Netherlands, New Zealand, Nigeria, Norway, Pakistan, Poland, Portugal, Russian Federation, Senegal, Slovakia, Spain, Sweden, Switzerland, Thailand, Tunisia, Turkey, Ukraine, United Kingdom, United States, Venezuela
MAXIMUM STRENGTH: 39,922 (38,614 troops and support personnel, 637 military observers, 671 civilian police and 4,058 staff (1994)
STRENGTH AT WITHDRAWAL: 2,675 FATALITIES: 207 CLASP(S): None

9. UNTAC - UNITED NATIONS TRANSITIONAL AUTHORITY IN CAMBODIA
LOCATION: Cambodia DATES: Mar. 1992 to Sept. 1993
COUNTRIES PARTICIPATING: (46) Algeria, Argentina, Australia, Austria, Bangladesh, Belgium, Brunei, Bulgaria, Cameroon, Canada, Chile, China, Colombia, Egypt, Fiji, France, Germany, Ghana, Hungary, India, Indonesia, Ireland, Italy, Japan, Jordan, Kenya, Malaysia, Morocco, Namibia, Nepal, Netherlands, New Zealand, Nigeria, Norway, Pakistan, Philippines, Poland, Russian Federation, Senegal, Singapore, Sweden, Thailand, Tunisia, United Kingdom, United States, Uruguay
MAXIMUM STRENGTH: 19,350 military and civilian personnel (1993)
STRENGTH AT WITHDRAWAL: 2,500 (approx.) FATALITIES: 78 CLASP(S): **UNAMIC** (later withdrawn)

10. UNOSOM II - United Nations Operation in Somalia II
COUNTRY/LOCATION: Somalia DATES: May 1993 to March 1995
COUNTRIES PARTICIPATING: (34) Australia, Bangladesh, Belgium, Botswana, Canada, Egypt, France, Germany, Ghana, Greece, India, Indonesia, Ireland, Italy, Korean Republic, Kuwait, Malaysia, Morocco, Nepal, Netherlands, New Zealand, Nigeria, Norway, Pakistan, Philippines, Romania, Saudi Arabia, Sweden, Tunisia, Turkey, United Arab Emirates, United States, Zambia, Zimbabwe
MAXIMUM STRENGTH: 30,800 authorized (28,000 military personnel and approximately 2,800 civilian staff)
STRENGTH AT WITHDRAWAL:14,968 FATALITIES: 147 CLASP(S): None

11. UNMIH - UNITED NATIONS MISSION IN HAITI
COUNTRY/LOCATION: Haiti DATES: Sept 1993 to June 1996
COUNTRIES PARTICIPATING: (34) Algeria, Antigua and Barbuda, Argentina, Austria, Bahamas, Bangladesh, Barbados, Belize, Benin, Canada, Djibouti, France, Guatemala, Guinea-Bissau, Guyana, Honduras, India, Ireland, Jamaica, Jordan, Mali. Nepal, Netherlands, New Zealand, Pakistan, Philippines, Russian Federation, St.Kitts & Nevis, St.Lucia, Suriname, Togo, Trinidad and Tobago, Tunisia, United States
MAXIMUM STRENGTH: 6,065 military personnel and 847 civilian police (1995)
STRENGTH AT TRANSITION TO UNSMIH: 1,200 troops and 300 civilian police FATALITIES: 6
CLASP(S): None

UNSSM - United Nations Special Service Medal
BACKGROUND: Established in 1994 by the Secretary-General of the United Nations, the United Nations Medal for Special Service is awarded to military and civilian police personnel serving the United Nations in capacities other than established peacekeeping missions or those permanently assigned to United Nations Headquarters. The Medal for Special Service may be awarded to eligible personnel serving for a minimum of ninety (90) consecutive days under the control of the United Nations in operations or offices for which no other United Nations award is authorized. Posthumous awards may be granted to personnel otherwise eligible for the medal who die while serving under the United Nations before completing the required 90 consecutive days of service.
CLASP(S): Clasps engraved with the name of the country or the United Nations organization (e.g., UNHCR, UNSCOM, MINUGUA, etc.) are added to the medal suspension ribbon and ribbon bar.

NATO MEDAL (KOSOVO)

Bronze

Instituted: 1999 (effective 13 Oct 1998)
Criteria: 30 days service in or 90 days outside the former Republic of Yugoslavia and the Adriatic Sea under NATO command in direct support of NATO operations
Devices: Bronze star
Notes: Qualifying service time includes dates between 13 Oct 1998 and a termination date to be announced.
KOSOVO Bar is not authorized for wear by U.S. military personnel.

Awarded by The North Atlantic Treaty Organization (NATO) to military personnel who have served under NATO command or in direct support of NATO operations in Kosovo.

The NATO Medal is a non-U.S. service award. The medal is awarded to members of the U.S. Armed Forces who served 30 days continuous or accumulated time under NATO command in direct support, of NATO operations on land, at sea, or in the air space of Kosovo and other territories of the Federal Republic of Yugoslavia, Albania, the former Yugoslav Republic of Macedonia and the Adriatic and Ionian Seas between 13 October 1998 and a termination date to be announced. Aircrews in Operation ALLIED FORCE between 24 March and 10 June 1999 will be deemed to have qualified after 15 sorties, regardless of time served. Personnel who served in direct support for 90 days in the adjacent areas of Italy, Greece and Hungary are also eligible.

The NATO Medal is a bronze disc with a raised edge. In the center of the medal is the NATO star surrounded by a wreath of olive leaves. The reverse of the medal contains the raised inscription IN SERVICE OF PEACE AND FREEDOM in English (on top) and in French (below). Around the edge is a raised double ring with the enclosed inscription NORTH ATLANTIC TREATY ORGANIZATION in English (on top) and in French (below). The ribbon is medium blue with a white center stripe and narrow white borders. The medal is sometimes issued with a clasp, which it is not worn by U.S. Armed Forces. Only one NATO Medal/Ribbon is authorized for wear. Should an individual become eligible for an additional NATO Medal, it will be indicated by a three sixteenth-inch bronze star on the first NATO Medal awarded.

MULTINATIONAL FORCE and OBSERVERS MEDAL

Bronze Numeral

Instituted: 1982
Criteria: 6 months service with the Multinational Force & Observers peacekeeping force in the Sinai Desert
Devices: Bronze numeral
Notes: Above date denotes when award was authorized for wear by U.S. military personnel

For service with the Multinational Force and Observers peacekeeping force in the Sinai Desert. The Multinational Force and Observers was created to act as a buffer between Israel and Egypt in the Sinai Peninsula.

The Multinational Force and Observers Medal is considered the same as a Non-U.S. Service Award. The award was established on 24 March 1982. Approval for US personnel to wear this decoration was granted by the Department of Defense on 26 July 1982. The medal was first awarded to individuals who had served at least 90 days with the Multinational Force and Observers. The time of service was increased to 170 days minimum after 15 March 1985.

The Multinational Force and Observers Medal is a bronze disc depicting a dove and an olive branch centered within two raised rings. Between the two rings is the raised inscription MULTINATIONAL FORCE at the top, and OBSERVERS at the bottom. There is a fine grid pattern in the background. The medal is suspended from a bronze rectangular bar, which is attached to the ribbon. The reverse of the medal has the raised inscription UNITED IN SERVICE FOR PEACE set in five straight lines. The ribbon is orange with a wide white center stripe flanked on each side with narrow olive green stripes. Bronze numerals are authorized for subsequent awards.

INTER-AMERICAN DEFENSE BOARD MEDAL

Gold

Instituted: 1981
Criteria: Service with the Inter-American Defense Board for at least one year
Devices: Gold star
Notes: Above date denotes when award was authorized for wear by U.S. military personnel

For service of a minimum of one year on the Inter-American Defense Board. The Inter-American Defense Board Medal is a non-U.S. service award. The medal was authorized by the Inter-American Defense Board on 11 December 1945 and authorized for wear by the Armed Forces of the United States by the Department of Defense on 12 May 1981. The criteria for the award is a minimum of one year service on the board in specific responsibilities.

The Inter-American Defense Board Medal is a bronze disc with a map projection of the Western Hemisphere in the center. Around the projection are the flags of the Nations of North and South America.

The reverse of the medal is blank, but when presented it is engraved with the recipient's name and the words FROM THE INTER-AMERICAN DEFENSE BOARD and FOR SERVICE. The ribbon is five equal stripes of red, white, blue, yellow and green. An additional period of service on the Inter-American Defense Board is denoted by a five-sixteenth inch gold star.

REPUBLIC OF VIETNAM CAMPAIGN MEDAL

Silver Date Bar

Instituted: 1966
Criteria: 6 months service in the Republic of Vietnam between 1961 and 1973 or if wounded, captured or killed in action during the above period
Devices: Silver date bar
Notes: Bar inscribed "1960- " is the only authorized version

Awarded by the Republic of Vietnam to members of the U.S. Armed Forces who served for six months in Vietnam.

The Republic of Vietnam Campaign Medal is a non-U.S. Service Award. The medal was authorized on 20 July, 1966 and amended on 31 January 1974. This campaign medal was awarded by the Republic of Vietnam to members of the U.S Armed Forces who served a minimum of six months in the Republic of Vietnam between 1 March 1961 and 28 March 1973, or who had provided direct combat support to the RVN during the period of the award. The Government of Vietnam awards the ribbon, but the medal must be purchased by recipients. Anyone qualifying for the Vietnam Service Medal was automatically awarded this medal.

The medal is a six pointed white enamel star, bordered in gold, with cut lined broad gold star points between. In the center of the star is a green disc, bordered in gold, with a map of Vietnam in gold and a red enamel flame. The reverse of the medal has a raised inscription VIETNAM in a lined circle with CHIEN-DICH at the top and BOI-TINH below separated by short lines. The ribbon is green with a white center stripe and white stripes near the edges. A silver scroll is attached to the ribbon with the date 1960 and a dash.

KUWAIT LIBERATION MEDAL (SAUDI ARABIA)

Gold Palm Tree

Instituted: 1991
Criteria: Participation in, or support of, Operation Desert Storm (1990-91)
Devices: Gold palm tree device
Notes: Support must have been performed in theater (e.g.: Persian Gulf, Red Sea, Iraq, Kuwait, Saudi Arabia, Gulf of Oman, etc.)

Awarded by the Kingdom of Saudi Arabia for participation in Operation Desert Storm.

The Kuwait Liberation Medal (Saudi Arabia) is a non-U.S. service award. The medal was authorized for wear by members of the U.S. Armed Forces by the Department of Defense on 7 October 1991. This medal was awarded by the Kingdom of Saudi Arabia to their troops and troops of the United Nations Coalition who served in the Kuwait war zone from 17 January to 28 February 1991.

The Kuwait Liberation Medal (Saudi Arabia) was designed by the Swiss firm of Huguenin Medailleurs. The medal consists of a silver star with fifteen large rays and fifteen small rays. In the center of the star is a disc with a gold map of Kuwait on a silver globe surrounded by two palm branches. Above the globe is a royal crown and above that are crossed swords and a palm tree, which is the emblem of Saudi Arabia. At the bottom of the design is a scroll with the raised inscription LIBERATION OF KUWAIT in Arabic and in English. The ribbon has a dark green center stripe flanked on either side by a narrow white stripe and a thin black stripe. The ribbon is edged with a narrow red stripe. The ribbon bar employs the emblem of Saudi Arabia (crossed swords and a palm tree) in the center.

KUWAITI LIBERATION MEDAL (EMIRATE OF KUWAIT)

Instituted: 1995
Criteria: Participation in, or support of, Operations Desert Shield and/or Desert Storm
Devices: None
Notes: Above date denotes when award was authorized for wear by U.S. military personnel

Awarded by the Kuwaiti Government for participation in Operation Desert Shield and Desert Storm.

The Kuwaiti Liberation Medal (Emirate of Kuwait) is a foreign award. The medal was authorized for wear by member of the U.S. Armed Forces by the Department of Defense in March 1995. This medal was awarded by the Kuwaiti Government for service in Operation Desert Shield and Desert Storm during the period from 2 August 1990 to 31 August 1993. Personnel must have served a minimum of one day on the ground, one day at sea, one day in aerial operations, or temporary duty of 30 days consecutive/60 days non-consecutive in support of military operations in the area.

The Kuwaiti Liberation Medal (Emirate of Kuwait) is a bronze disc with the Coat of Arms of the State of Kuwait. The Coat of Arms consists of the Shield of Kuwait in black, red, white and green enamel superimposed on a falcon with wings spread. Between the falcon's upspread wings is a disk containing a sailing ship with the name of state above it. At the top of the medal is a raised Arabic inscription 1991 LIBERATION MEDAL. The medal is suspended by a wreath attached to a bronze rectangle which is attached to the ribbon. The reverse of the medal has a map of Kuwait over a background of rays. The ribbon has three wide vertical stripes of red, white and green below a wide horizontal trapezoidal-shaped black stripe.

REPUBLIC OF KOREA WAR SERVICE MEDAL

Instituted: 1951; approved for U. S. Veterans in 1999
Criteria: Service in Korean War theater for 30 consecutive days, or 60 nonconsecutive days
Devices: None*

* Some original 1953 medals had a tae-guk in the center of the drape like the ribbon.

Awarded by the Republic of Korea for service in the Korean War. The Republic of Korea War Service Medal is a foreign award recently approved for award to U. S. Veterans by Congress. It was awarded to members of the United Nations Command who served 30 consecutive days, or 60 nonconsecutive days in Korea or its territorial waters. Aircrews who flew over Korea in combat, or support operations are also eligible.

The Republic of Korea War Service Medal is a bronze disc with a raised edge. The medallion has the United Nations Emblem over crossed artillery shells; both surrounded by olive leaves. The ribbon has a yellow border on each side with narrow light blue and white stripes; a thin red stripe is centered on the white stripe. In the center of the ribbon and earlier medal drapes (1950's) is an ancient oriental symbol called a tae-guk (the top half is red and the bottom half is blue).

Marksmanship Badges

In 1910 the United States Navy established the Navy Sharpshooters Medal, which was replaced by the Navy Expert Rifleman Medal in 1920. These medals were designated badges, but were designed like medals and are worn as decorations and medals.

Navy Expert Rifleman Badge

Navy Expert Pistol Shot Badge

The Navy Expert Rifleman Badge - was designed by the U. S. Mint and is awarded to members of the U. S. Navy and Naval Reserve who qualify as expert with a rifle on a prescribed military rifle course. The medallion is a bronze disc bordered with a rope edge. There is a smaller disc superimposed at the top, which is attached to a ribbon. The lager disc has a raised "bull's eye" in the center. Above the "bull's eye" is a raised inscription EXPERT RIFLEMAN; around the lower edge, UNITED STATES NAVY. The smaller disc contains the seal of the United States Navy. The reverse of the medallion is blank for engraving. The ribbon is navy blue with three thin light green stripes.

The Navy Expert Pistol Shot Badge - was created at the same time as the Navy Expert Rifleman Badge and is awarded to Naval Personnel who qualify as experts on a prescribed military course. The medallion is the same as the Expert Rifleman badge except for the raised inscription EXPERT PISTOL SHOT. The ribbon is navy blue with two narrow light green stripes at each edge.

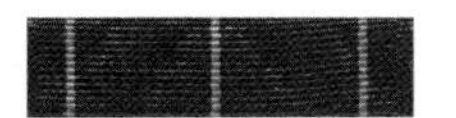

Navy Rifle Marksman Ribbon

Navy Pistol Marksman Ribbon

The Navy Rifle Marksman Ribbon - was authorized by the Secretary of the Navy on 14 October 1969. The Navy Expert Rifleman Medal and Ribbon were re-designated by awarding the ribbon bar to Marksman and Sharpshooters and ribbon bar and medal to those who fired Expert. The Marksman Ribbon is unadorned; the Sharpshooter Ribbon has a bronze letter "S" and the Expert Ribbon a silver "E." The ribbon is navy blue with three thin stripes of light green. Attachment letters are affixed to the center of the ribbon.

The Navy Pistol Marksman Ribbon - was authorized by the Secretary of the Navy on 14 October 1969. The Navy Expert Pistol Shot Medal and Ribbon were re-designated by awarding the ribbon bar to Marksman and Sharpshooters and ribbon bar and medal to those who fired Expert. The Marksman Ribbon is unadorned; the Sharpshooter Ribbon has a bronze letter "S" and the Expert Ribbon a silver letter "E." The ribbon is navy blue with two thin stripes of light green. Attachment letters are affixed to the center of the ribbon; the medal has no attachments.

Commemorative Medals

The United States Government, State Governments, Veterans Organizations, private mints and individuals have a long history of striking Commemorative Medals to recognize and honor specify military victories, historical events and military service to the United States.

The tradition of honoring U.S. military heroes began when the Continental Congress awarded gold and silver medals to our victorious commanders during the Revolutionary War. While these medals were struck as table display medals, General Gates, the victor of Saratoga, chose to wear his medal from a neck ribbon for his official portrait.

Naval heroes were also honored. Captain John Paul Jones received this gold medal for his victorious engagement with HMS Serapes in 1779.

These first Congressionally authorized medals were the forerunners of modern combat decorations.

John Paul Jones Medal

Veteran's organizations have also issued medals to their members commemorating their service. Some of the more significant examples are the Grand Army of the Republic medals worn by Union veterans at reunions following the Civil War and the Aztec Club Medal struck by veterans of the war with Mexico.

Grand Army of the Republic Reunion Medal

Aztec Club Medal

The U.S. Mint regularly produces commemorative medals to celebrate and honor individuals, places, events. The Vietnam Veterans National Medal commemorates the courage and dedication of the men and women who served in the Vietnam War. The Missing in Action Medal is a miniature replica of the 3-inch medal authorized for presentation to the next-of-kin of American personnel missing or otherwise unaccounted for in Southeast Asia.

Missing in Action Medal

The 200th anniversaries of the U.S. Army, Navy, Marine Corps, and Coast Guard were also celebrated with the striking of medals. The Persian Gulf National Medal honors Persian Gulf War veterans.

A complete listing of all commemorative medals available from the U.S. Mint can be obtained by calling (202) 283-2646.

While the Federal Government issues commemorative medals from the U.S. Mint, state and county governments use private mints and contractors to strike medals honoring veterans. Veterans associations such as the American Legion, Veterans of Foreign Wars, and the Daughters of the Confederacy have also issued commemorative medals.

Commemorative medals reflect the American spirit. The 75th Anniversary of World War I and the 50th Anniversary celebrations of both World War II and the Korean War have occasioned the need for commemorative medals to honor the veterans of those conflicts. The most recent example is the Cold War Victory Commemorative Medal struck to fill the void created when Congress authorized a Cold War Victory Recognition Certificate but elected not to fund a medal.

Cold War Victory Commemorative Medal

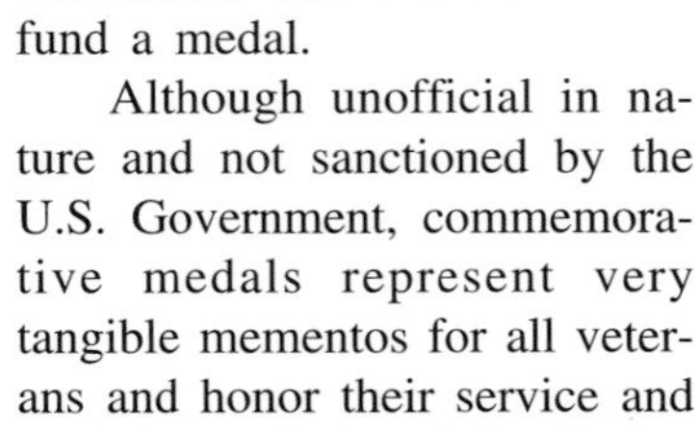

Although unofficial in nature and not sanctioned by the U.S. Government, commemorative medals represent very tangible mementos for all veterans and honor their service and sacrifice. Commemorative Medals are designed for display and not worn on military uniforms.

The commemorative medals shown on page 117 are representative of high quality, well-designed commemoratives struck to honor veterans for service during the past sixty years. These medals were struck by Medals of America, Inc.

The World War II Victory Commemorative Medal

The medal depicts The Goddess of Truth and Freedom raising the sword and laurel crown of Victory over crushed shields and swords of the Axis Powers. The scene is wreathed in a crown of laurel leaves, ancient symbol of victory and honor. The medal is struck to honor all who served in the United States Forces during WW II.

On the ribbon drape has a wide red stripe in the center and is bound by two thin white stripes bordered with two wide blue bands. There is a gold stripe in the center of the blue bands with a thin red stripe in the center of the gold.

The Korean Defense Commemorative Medal

The 50th Anniversary Korean Defense Commemorative Medal Period of Service is designed to honor all military and civilian personnel who have served in the defense of the Republic of Korea between 1950 and 2000.

The front of the medal depicts an armed American standing guard at the gate to South Korea. The words, "Korean Defense Commemorative" are across the top . The dates 1950, 2000 flank the gates while the legend, "Commemorating 50 Years in the Defense of South Korea" is in raised letters across the bottom.

The ribbon drape is United Nations blue and alternating stripes of white. There is a thin red line in the center.

The Republic of Vietnam Defense Commemorative Medal

Struck to honor all military and civilian personnel who served in the defense of South Vietnam in country or in support of operations either off shore or in adjacent countries (i.e. Thailand, Guam etc.) between 1960 and 1975.

The Medal depicts a coiled Dragon representing Vietnam bordered by bamboo. The Words "Republic of Vietnam Defense, 1960-1975" are over the Dragon while the words "Commemorating Service in the Defense of Vietnam" are at the bottom.

The ribbon drape is green with yellow borders and a center band of yellow and red stipes reflecting the Republic of Vietnam's National Flag.

The Cold War Commemorative Medal

The Cold War Victory Commemorative medal was inspired by the Cold War Certificate of Recognition created by Congressional Resolution to recognizes members of the Armed Forces who served during the Cold War between September 2, 1945 and December 26, 1991.

The medal depicts an American Eagle holding arrows in the right claw and the olive branch in the left (the war eagle). The inscription above reads, "Duty, Honor, Country." "Cold War Victory Commemorative" is in raised letters across the top of the medal and 13 raised stars are embossed along the bottom edge.

The ribbon is light grey; the center has stripes of red, white and blue.

The Combat Service Commemorative Medal

Struck to honor all Soldiers, Sailors, Marines, Airmen and Coast Guard Personnel who served in an overseas combat theater or in expeditionary combat operations.

The medal is bronze and is 1-1/4 inches in diameter. The front of the medal depicts key symbols representing the four branches of the Armed Forces. The words "Combat Service" are over the three spears representing air, land and sea forces. Beneath the spear heads are pilot's wings over body armor with crossed cannon and rifle, all held together on the arms and flukes of an anchor. The word "Commemorative" is at the bottom separated from the words "Combat Service" by 13 stars representing the original colonies.

The ribbon is designed to reflect the National colors separated by a broad band of gold.

The Overseas Service Commemorative Medal

Struck to honor all Soldiers, Sailors, Marines, Airmen and Coast Guard personnel who served overseas or in expeditionary operations for 30 days or more.

The Medal depicts an American Eagle with the national shield over looking two globes showing both sides of the globe. Beneath the globes are five stars representing the branches of the U.S. Armed Forces. Around the top of the medal are the words "Overseas Service ." The word "Commemorative" is at the bottom with laurel leaves flanking the globes.

The Ribbon colors were chosen to represent land, sea and air, with the national colors running down the center.

Displaying Awards

In the United States, it is quite rare for an individual to wear full-size medals once no longer on active duty. Unfortunately, many veterans return to civilian life with little concern for the state of their awards. In the euphoria of the moment, medals are tucked away in corners or children play with them, often causing irreparable damage to these noble mementos of a man's or woman's patriotic deeds. The loss or damage of these medals is sad, since awards reflect the veteran's part in American History and are totally unique and personal to each family.

The most appropriate use of military medals after active service is to mount the medals for permanent display in home or office. This reflects the individual's patriotism and the service rendered the United States. Unfortunately, there are very few first class companies in the United States which possess the expertise to properly prepare and mount awards and other personal militaria. The following pages provide examples of the formats, mounting methods and configurations employed by Medals of America in Fountain Inn, South Carolina to display military decorations. The examples range from World War II, Korea, Vietnam, Kuwait and Kosovo to peacetime service.

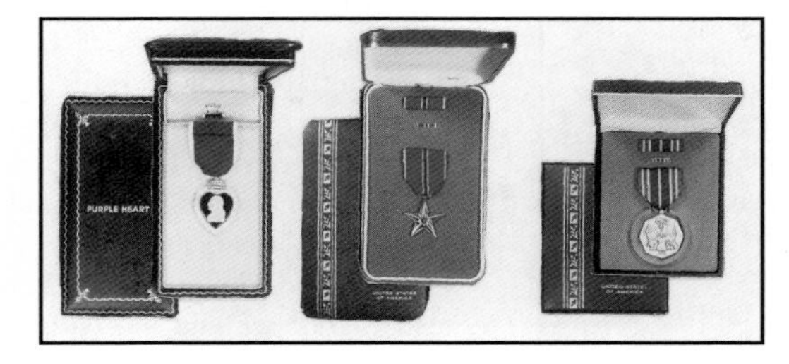

Decorations are usually awarded in a presentation set which normally consists of a medal, ribbon bar and lapel pin, all contained in a special case. During World War II, the name of the decoration was stamped in gold on the front of the case. However, as budget considerations assumed greater importance, this practice was gradually phased out and replaced by a standard case with "United States of America" emblazoned on the front.

At the present time, the more common decorations, (e.g., Achievement and Commendation Medals), come in small plastic cases, suitable only for initial presentation and storage of the medal. Using this case in its open position for prolonged display exposes the entire presentation set to dust, acids and other atmospheric contaminants which can cause tarnish and/or serious discoloration.

Outside the case, medals and ribbons should be handled as little as possible, since oils and dirt on the hands can cause oxidation on the pendant and staining of the ribbon.

The most effective method of protecting awards involves the use of a shadow box or glass display case with at least 1/2 inch between the medals and the glass. This provides a three dimensional view and protects the medal display in a dust-free environment. Cases which press the medals and ribbons against the glass can disfigure the ribbon, cause discoloration and, in some extreme cases, damage the medal.

The greatest mistake an ordinary frame shop can make is in the actual process of mounting the medals. They often clip off prongs or pins on the back of a medal to ease the task of gluing the medal to a flat surface. The physical alteration destroys the integrity of the medal and the use of glues ruins the back of the ribbon and medal. The net result, ignoring the intrinsic value of the piece, is serious damage to a valued heirloom and keepsake.

The best way to mount medals is in a wooden case especially designed for that purpose. They can be obtained either with a fold-out easel back for placement on a table, desk or mantle, or with a notched hook on the rear of the wooden frame for hanging on a wall. The case should also have brass turnbuckles on the back to facilitate removal of the mounting board for close examination of the medals or rearrangement of the displayed items.

The mounting board is absolutely critical. Velvet, flannel and old uniforms, just don't do the job. A first class mounting system starts with acid-free Befang or Gator board at least 1/4 inch thick. This board is very sturdy, being composed of two layers of hard, white paper board sandwiched around a foam core. Over this, a high quality velour-type material to which velcro will adhere is glued and pressed down evenly. The medals are mounted using velcro tape; one piece over the ribbon mounting pin and one, about the size of a nickel, on the medal back. The velcro locks the medal very firmly into place without causing any damage and alleviates the need to cut off the pin backs. Badges or ribbon bars with pin backs can be mounted by pressing the prongs through the fabric into the gator board, which moulds around the prongs. A little velcro tape on the back of the pronged device adds extra holding power.

Patches, brass plates, dog tags and other mementos can easily be added this way. The great beauty of this method is found not only in its eye appeal, but also that one can add to the display or rearrange the existing contents by gently peeling the medal off as simply as opening a velcro zipper. Prong devices can also be moved easily, since the foam core closes in behind the prong as it is removed, thus effectively sealing the hole once more.

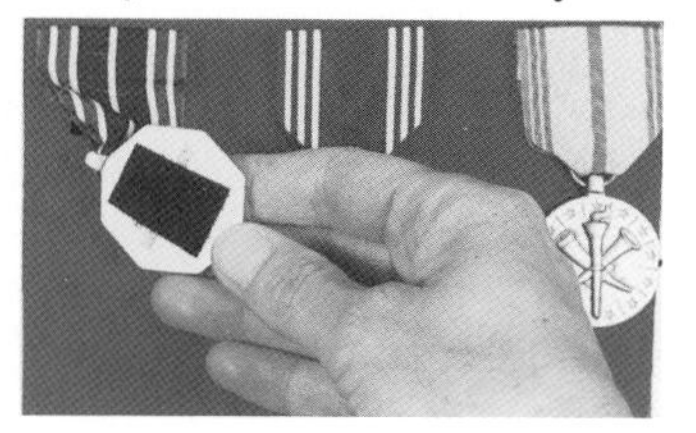

The final element in the process is the frame itself. While oak and other heavy woods make very handsome pieces of furniture, they are not a good choice for a frame. The frame's weight puts a great strain on modern plasterboard walls when an extensive medal display is attached via standard hooks and nails. In addition, handling a heavy frame by very young or very old hands increases the chance it could be accidentally dropped. For these reasons, frames should be milled from a lightweight wood with good staining characteristics. Bass wood is considered the best for the purpose and some poplar is acceptable. Metal frames, on the other hand, should be avoided, owing to their heavy weight and to the bright coloring which can conflict with the patina of the medals. Finally, the wood stain, (e.g., walnut stain), should reflect a rich, warm glow to properly envelop and enhance the medal display.

JOHN M. DORSEY
U.S.S. RANDOLPH (CVS-15)
1961-1963

An essential part of the display case is the brass plate which provides the key information pertaining to the recipient of the displayed awards. This is definitely a place where one should not cut corners. A bargain basement brass or gold colored plastic nameplate will cheapen and distract from an otherwise elegant display case. Conversely, a high quality brass plate with good quality engraving will forever enhance the dignity of the medal display.

Whenever possible, the engraved letters should be blackened to provide enhanced contrast and visibility. The plate should, as a minimum, contain the full name, assigned unit and time frame. Other useful items are rank, service number and branch of the service if space permits. The contents of a nameplate are obviously a personal preference, but experience has shown that a limit of four or five lines can enhance and compliment the display, while greater numbers are a distraction.

On page 71 are some examples of different display cases. The Gulf War case displayed here has an enlisted Surface Warfare Insignia at the top with a set of ten ribbons underneath. Four ribbons are ribbon only awards; they are Presidential Unit Citation, Navy Unit Commendation, Navy Meritorious Unit Commendation and Navy Sea Service Deployment Ribbon. The medals are from left to right: Navy and Marine Corps Achievement Medal, Navy Good Conduct Medal, National Defense Service Medal, Southwest Asia Service Medal, Kuwait Liberation Medal (Saudi Arabia) and Kuwait Liberation Medal (Emirate of Kuwait). Metal rating badges flank the engraved nameplate.

Bibliography

Baldwin, H.W. - *What You Should Know About The Navy*

Belden, B.L. - *United States War Medals*, 1916

Borthwick, D. and Britton, J. - *Medals, Military and Civilian of the United States*, 1984

Borts, L.H. and Foster, F.C. - *United States Military Medals 1939 to Present,* 1998

Borts, L.H. - *United Nations Medals and Missions,* 1997

Department of the Navy - *United States Navy Uniform Regulations,* 1998

Dorling, H.T. - *Ribbons and Medals,* 1983

Fowler, W. and Kerrigan, E. - *American Military Insignia and Decorations,* 1995

Gleim, A.F. - *United States Medals of Honor,* 1862-1989, 1989

Kerrigan, E. - *American Badges and Insignia,* 1967

Kerrigan, E. - *American Medals and Decorations,* 1990

Kerrigan, E. - *American War Medals and Decorations,* 1971

Kerrigan, E. - *Guidebook of U.S. Medals,* 1994

McDowell, C.P. - *Military and Naval Decorations of the United States*

National Geographic Society - *Insignia and Decorations of the United States,* 1944

Rankin, R.H. - *Uniforms of the Sea Services*

Rosignoli, G. - *Badges and Insignia of World War II*

Rosignoli, G. - *The Illustrated Encyclopedia of Military Insignia of the 20th Century*

Smith, Richard W. - *Shoulder Sleeve Insignia of the U.S. Armed Forces 1941-1945,* 1981

Stacey, John A. - *U. S. Navy Rating Badges 1885 - 1996,* 1986 and 1996

Standberg, J.E. and Bender, R.J. - *The Call to Duty,* 1994

Sylvester, J. and Foster, F.C. - *The Decorations and Medals of the Republic of Vietnam and Her Allies,* 1995

Thompson, James G. - *Decorations, Medals, Ribbons, Badges and Insignia of the United States Marine Corps,* 1998

United States Naval Institute - *The Blue Jackets Manual United States Navy,* Tenth Edtion

U.S. Navy Instruction SECNAVINST 1650.1F - *Navy and Marine Corps Awards Manual,* 1991

U.S. Navy Manual NAVPERS 15,790 - *Decorations, Medals, Ribbons and Badges of the United States Navy, Marine Corps and Coast Guard, 1861-1948,* 1 July 1950

U. S. Navy - *United States Navy Uniform Regulations*

U. S. Navy - *All Hands*

Vietnam Council on Foreign Relations - *Awards and Decorations of Vietnam,* 1972

Windas, Cedric W. - *Traditions of the Navy,* 1942 and 1954

INDEX

A

About the Author 2
Acknowledgments 2
Advanced Undersea Weaponsman 32
Aerographer 18
Aerographer's Mate 23
Aiguillettes 44
Air Medal 82
Air Traffic Control Technician 18
Air Traffic Controller 23
Air Traffic Controlman 23
Aircraft Gunner 32
Aircrew Survival Equipmentman 28
Aircrewman 32
Airship Insignia 32
Airship Rigger 23
American Campaign Medal 91, 93
American Defense Service Medal 91, 93
Amphibious Insignia 32
Andre Medal 9
Antarctic Service Medal 97
Anti-Aircraft Machine Gunner 32
Apprentice Training Graduates 22
Armed Forces Expeditionary Medal 98
Armed Forces Reserve Medal 104
Armed Forces Service Medal 101
Armed Guard 32
Asiatic-Pacific Campaign Medal 93
Assault Boat Coxswain 32
Attachments and Devices 48, 49, 50, 51
Aviation Antisubmarine Warfare Technician 23
Aviation Boatswain 18
Aviation Boatwain's Mate 23
Aviation Electrician's Mate 23
Aviation Electronics Technician 18, 23, 26
Aviation Electronicsman 23
Aviation Experimental Psychologist and Aviation Physiologist 36
Aviation Fire Control Technician 23
Aviation General Utility 32
Aviation Guided Missilman 23
Aviation Machinist's Mate 23
Aviation Maintenance Administrationman 24
Aviation Maintenance Technician 18
Aviation Operations Technician 18
Aviation Ordnance Technician 18
Aviation Ordnanceman 24
Aviation Photographer's Mate 24
Aviation Pilot 24
Aviation Radio Technician 24
Aviation Radioman 23, 24
Aviation Storekeeper 24
Aviation Structural Mechanic (Aviation Metalsmith) 24
Aviation Support Equipment Technician 24
Aviation Warfare Specialist Insignia 36
Award Certificates 52
Award Displays 71
Awards and Decorations 46
Awards Displays 118

B

Background and History 6
Badge of Military Merit 9
Badges and Insignia 7
Baker 24, 31
Balloon Pilot Insignia 35
Bandmaster 27
Basic Explosive Ordnance Disposal Warfare Insignia 39
Basic Parachutist Insignia 38
Bibliography 120
Blacksmith 24
Board of Admiralty Seal 11
Boatswain 18
Boatswain's Mate 24, 25
Boatswain's Pipe and Lanyard 45
Boiler Technician 24
Boilermaker 24
Boilerman 24
Bombsite Mechanic 32
Brassards 45
Breast Insignia 35, 58, 59, 60
Bronze Star Medal 79
Buglemaster 24
Bugler 24
Builder 25
Buttons 45

C

Cap Insignia 12
Carpenter's Mate 25, 28
Certificates 52
Chaplain Corps (Buddist) 17
Chaplain Corps (Christian) 17
Chaplain Corps (Jewish) 17
Chaplain Corps (Muslim) 17
Chief Commissary Steward 25
Chief Petty Officer 21
China Service Medal 90, 91, 93
Civil Engineer Corps 17, 19
Civilian Dress 55
Claiming Medals 53
Color Plates 56
Commissaryman 29
Combat Action Ribbon 85
Combat Aircrew Insignia 36
Command Ashore/Project Manager Insignia 38
Command-at-Sea Insignia 37
Commemorative Medals 116, 117
Commissaryman 24, 25, 31
Commodore 7
Communications Technician 18, 25
Construction Battalion 32
Construction Electrician 25
Construction Electrician's Mate 25
Construction Mechanic 25
Cook 25, 29, 33
Cook 31
Coppersmith 25
Coxswain 25
Craftmaster Insignia 38
Cryptologic Technician 18, 25

D

Damage Controlman 25, 28, 29
Data Processing Technician 18, 25
Data Systems Technician 25
Decoration 46, 47
Decorations, Medals and Ribbons 9
Deep Submergence Insignia 39
Defense Distinguished Service Medal 66, 75
Defense Meritorious Service Medal .. 81
Defense Superior Service Medal 77
Dental Corps 17
Dental Technician 25
Devices 48, 49, 50, 51
Devices on Medals 51
Devices on Ribbons 50
Disbursing Clerk 25
Displaying Awards 118
Distinguished Flying Cross 78
Distinguishing Marks 7, 32
Diver 1st Class 33
Diver 2nd Class 33, 39
Diver Salvage 33
Diving (Medical) Insignia 38
Diving Medical Technician Insignia .39
Diving Officer 18, 38
Draftsman 26

E

Electrician's Mate 25, 26, 28
Electronics Technician ... 19, 25, 26, 29
Electronics Technician's Mate 26, 29
Electronics Warfare Technician 26
Engineering Aid 26, 31
Engineering Technician 19
Engineman 26, 27
Enlisted Rank/Rate Insignia 20
European-African-Middle Eastern Campaign Medal 92
Ex-Apprentice Mark 33
Expert Lookout 33
Expert Rifleman and Pistol Shot 34
Explosive Ordnance Disposal Technician 19, 33

F

Fire Control Radar Operator 33
Fire Control Technician 26
Fire Controlman 25, 26
Fire Fighter Assistant 33
First Class Diver Insignia 39
Fleet Marine Force Ribbon 89
Flight Nurse Insignia 36
Flight Surgeon Insignia 35
Foreign Decorations 69, 70
Formal Civilian Wear 55

G

Gas Turbine System Technician 26
Grade Insignia, Metal 16
Group Rate Marks 21
Guided Missileman 27
Gun Captain 33
Gun pointer 33
Gun Pointer First Class 33
Gunner's Mate 26, 31

H

Hospital Corpsman 26, 28
Hospital Steward 28

Hull Maintenance
Technician 26, 28, 29
Humanitarian Service Medal 101

I

Identification Badges 40, 61
Illustrator Draftsman 26
Information Technology
Specialist 25, 29
Instrumentman 26
Integrated Undersea Surveillance
System Insignia 39
Intelligence Specialist 26
Intelligence Technician 19
Inter-American Defense Board
Medal 113
Interior Communications
Electrician 26, 28
Introduction 5

J

John Paul Jones Medal 9
Joint Chiefs of Staff Identification
Badge (JCS ID) 41
Joint Meritorious Unit Aweard 85
Joint Service Achievement Medal 84
Joint Service Commendation Service
Medal 83
Joint/Unified Command Identification
Badges 43
Journalist 27
Judge Advocate General Corps 17

K

Kirk, Shelby Jean 3
Korean Service Medal 96
Kosovo Campaign Medal 100
Kuwait Liberation Medal
(Saudi Arabia) 113

L

Lapel Pin 47
Large Medals 54
Law Community 17
Legalman 27
Legion of Merit 77, 78
Line and Staff Corps Devices 63
Lithographer 27, 28

M

Machine Accountant 25
Machinery Repairman 27
Machinist 27
Machinist's Mate 24, 27
MailClerk 27
Mailman 27
Marksman 34
Marksmanship Awards 70
Marksmanship Badges 115
Master Chief Petty Officer 21
Master Chief Petty Officer
of the Navy 21
Master Diver 33
Master Diver Insignia 38
Master Explosive Ordnance Disposal
Warfare Insignia 39
Master Horizontal Bomber 33
Master-at-Arms 27
Mechanic 25
Medal Clasps 48
Medal for Humane Action 95
Medal of Honor 9, 65, 74
Medical Corps 17
Medical Service Corps 17
Meritorious Service Medal 81
Mess Management
Specialist 25, 27, 31
Metalsmith 27
Mine Assemblyman 33
Mine Warfare Insignia 33
Minecraft Personnel Shoulder
Patch 33
Mineman 27, 31
Miniature Medals 54
MINURSO - United Nations Mission for
the Referendum 110
Missile Technician 27
Molder 27, 29
Motor Machinist's Mate 26, 27
Multinational Force and Observer
Medal 112
Musician 27

N

National Defense Service Medal 95
NATO Award Certificate 52
NATO Medal 109
NATO Medal for Kosovo 112
Naval Aircrew Insignia 36
Naval Astronaut (NFO) Insignia 35
Naval Astronaut Insignia 35
Naval Aviation Observer and Flight
Meteorologist Insignia 35
Naval Aviation Observers (Radar)
Insignia 35
Naval Aviation Observers (Tactical)
Insignia 35
Naval Aviation Supply Corps
Insignia 35
Naval Aviator Insignia 35
Naval Flight Officer Insignia (NFO)
Insignia 36
Naval Parachutist Insignia 38
Naval Reserve Medal 104
Naval Reserve Merchant Marine
Insignia 39
Naval Reserve Meritorious Service
Medal 89
Naval Reserve Sea Service Ribbon 103
Navy and Marine Corps Achievement
Medal 84
Navy and Marine Corps Commendation
Medal 83
Navy and Marine Corps Medal 79
Navy and Marine Corps Overseas
Service Ribbon 103
Navy Arctic Service Ribbon 102
Navy Career Counselor Badge 43
Navy Counselor 27
Navy Cross 75
Navy Distinguished Service Medal 76
Navy E 33
Navy E Ribbon 86
Navy Expeditionary Medal 90
Navy Expert Pistol Shot Badge 115
Navy Expert Rifleman Badge 115
Navy Fleet/Force/Command Master
Chief Badges 42
Navy Good Conduct Medal 87, 88
Navy Insignia 11
Navy Master At Arms (MAA)/Law
Enforcement Badges 43
Navy Medal of Honor 9, 74
Navy Meritorious Unit Commendation
Ribbon 86
Navy Occupation Service Medal 94
Navy Pistol Marksman Ribbon 115
Navy Recruit Company Commander
Badge 42
Navy Recruit Training Service
Ribbon 104
Navy Recruiting Command Badge 42
Navy Recruiting Service Ribbon 103
Navy Rifle Marksman Ribbon 115
Navy Sea Service Deployment
Ribbon 102
Navy Uniform Regulations, Wearing
Ribbons and Medals 54
Navy Unit Commendation 86
Nuclear Weaponsman 27
Nurse Corps 17

O

Ocean Systems Technician 27
Office of the Secretary of Defense
Identification Badge 41
Officer Rank Insignia 14
Officer Cap Insignia 11
Officer's Steward 29, 31, 33
Operations Specialist 28
Operations Technician 19
Opticalman 28
Ordnance Battalion 33
Ordnance Technician 19
Outstanding Volunteer Service
Medal 102

P

Painter 25, 28
Parachute Rigger 23, 28, 34
Parachuteman 34
Patrol Torpedo Boat 34
Patternmaker 28
Pay Grade 20
Personnelman 28
Petty Officer 21
Pharmacist's Mate 26, 28
Philippine Defense Medal and
Ribbon 107
Philippine Independence
Medal 69, 108
Philippine Liberation Medal and
Ribbon 107
Philippine Republic Presidential Unit
Citation 105
Photographer 19, 28
Photographer's Mate 24, 28
Photographic Intelligenceman 28
Physician's Assistant 19
Pipefitter 28
Pistol Marksman Ribbon 115
Placement of Devices on Medals 51
Placement of Devices on Ribbons 50

Postal Clerk 28
Presidential Service Badge (PSB) 40
Presidential Unit Citation 85
Printer 28
Prisoner of War Medal 87
Purple Heart 9, 80

Q

Quartermaster 28, 29

R

Radarman 28
Radio Technician 26, 29
Radio Technician's Mate 24
Radioman 29, 31
Rank and Rate Insignia 62
Rank/Rate Insignia, Enlisted 20
Ranks, Rates and Ratings 13
Rating Badges 64
Religious/Program Specialist 29
Repair Technician 19
Republic of Korea Presidential Unit
  Citation 105
Republic of Vietnam Armed Forces
  Honor Medal 105
Republic of Vietnam Campaign
  Medal 113
Republic of Vietnam Civil Actions Unit
  Citation 106
Republic of Vietnam Gallantry Cross
  Unit Citation 106
Republic of Vietnam Presidential Unit
  Citation 106
Reserve Special Commendation
  Ribbon 88
Retired Personnel Lapel Button 12
Ribbon Devices 73
Ribbon Precedence Chart 72
Ribbons 54, 72, 73
Ribbons with Medals 54
Rifle and Pistol Qualification
  Marks 34
Rifle Marksman Ribbon 115
Right Breast Ribbons When Wearing
  Medals 49

S

Saudi Arabian Award Certificate 52
Scuba Diver Insignia 39
Seabee Combat Warfare Specialist
  Insignia 38
Seal of the Department of the Navy .. 11
Seal of the United States Navy 11
Seaman Gunner 34
Second Class Diver Insignia 39
Security Technician 19
Senior Chief Petty Officer 21
Senior Explosive Ordnance Disposal
  Warfare Insignia 39
Service Aiguillettes 44
Service Medals 67
Service Stripes 22, 64
Sharpshooter 34
Shipfitter 28, 29
Ship's Clerk 19
Ship's Cook 24, 25, 29, 31
Ship's Cook (Butchers) 25
Ship's Serviceman 29
Shoulder Boards 16, 64
Signalman 29
Silver Star 76
Sleeve Devices, Commissioned
  Officers 17
Sleeve Devices, Warrant Officers 18
Small Craft Insignia 38
Sonar Technician 29
Sonar Technician - Surface 27
Sonarman 29
Soundman 29
Southwest Asia Service Medal 100
Special Operations Insignia 38
Special Warfare Insignia 38
Specialist (A) Physical Training
  Instructor, Airship Rigger &
  Carburetor Mechanic 29
Specialist (B) Master At Arms,
  Stevedore 29
Specialist (C) Classification Interviewer,
  Chaplain's Assistant 29
Specialist (E) Recreation Assistant,
  Physical Training Instructor 29
Specialist (F) Fire Fighter 30
Specialist (G) Aviation Free, Anti-aircraft
  and Gunnery Instructors 30
Specialist (H) Harbor Defense
  Sonarman 30
Specialist (I) IBM Operator, Punched
  Card Account Machine Operator .. 30
Specialist (K) Chemical Warfareman,
  Telecommunication Censorship
  Technician 30
Specialist (M) Mail Clerk, Underwater
  Mechanic 30
Specialist (O) Inspector of Naval
  Material 30
Specialist (P) Photographer,
  Photographic Specialist 30
Specialist (Q) Communication Security,
  Communication Specialist 30
Specialist (R) Recruiter,
  Transportationman 30
Specialist (S) Entertainer, Shore Patrol
  and Security MAA 30
Specialist (T) Teacher, Instructor,
  Transportation Airman 30
Specialist (U) Utility 30
Specialist (V) Transport Airman,
  Aviation Pilot 30
Specialist (W) Welfare (Chaplain's
  Assistant), Recreation Leader 30
Specialist (X) Specialist Not Elsewhere
  Classified 30
Specialist (Y) Control Tower
  Operator 31
Specialty Marks 23
SSBN Deterrent Patrol Insignia 37
Stacey, John A. 2
Steelworker 31
Steward 31
Storekeeper 31
Storekeeper (D) Dispersing 25
Strikers 22
Submarine Combat Patrol Insignia ... 37
Submarine Engineering Duty
  Insignia 36
Submarine Insignia 34, 36
Submarine Medical Insignia 36
Submarine Supply Corps Insignia 36
Supply Corps 17
Surface Warfare Dental Corps
  Insignia 37
Surface Warfare Insignia 37
Surface Warfare Medical Corps
  Insignia 37
Surface Warfare Medical Service Corps
  Insignia 37
Surface Warfare Nurse Corps
  Insignia 37
Surface Warfare Supply Corps
  Insignia 37
Surveyer 26, 31

T

Table of Contents 4
Technical Nurse Warrant Officer 19
Teleman 27, 28, 31
Torpedoman 31
Torpedoman's Mate 31
Tradesman (Training Devices Man) ... 31
Turret Captain 31
Types of Medals and Ribbons 47

U

U.S. Personal Decorations 66, 67
U.S. Service Medals 68, 69
UNAMIC - United Nations Advance
  Mission in Cambodia 111
Underwater Demolition Team Insignia 38
Underwater Mechanic 31
Underwater Ordnance Technician 19
Uniform Insignia 57
Unit Identification Marks (UIM's) 22
United Nations (Korean) Service
  Medal 108
United Nations (Observer) Medal ... 109
United Nations Medals ... 109, 110, 111
United Nations Missions 110
United States Antarctic Expedition
  Medal 97
UNPROFOR - United Nations Protection
  Force 111
UNSF - United Nations Security Force in
  West New Guinea 110
UNTSO - United Nations Truce
  Supervision Organization 110
Utilities Man 31

V

Vice Presidential Service Badge
  (VPSB) 40
Vietnam Service Medal 99

W

Warren, Thomas W. 1
Weapons Technician 31
Wearing of Medals, Insignia and the
  Uniform by Veterans
  and Retirees 55
Wearing Ribbons and Medals 54
World War II Victory Medal 94

Y

Yeoman 31